ADSP-2100 FAMILY USER'S MANUAL

ANALOG DEVICES TECHNICAL REFERENCE BOOKS

Published by Prentice Hall
   Analog-Digital Conversion Handbook
   Digital Signal Processing in VLSI
   Digital Signal Processing Applications Using the ADSP-2100 Family
   Digital Signal Processing Laboratory Using the ADSP-2101 Microcomputer

Published by Analog Devices
   Nonlinear Circuits Handbook
   Transducer Interfacing Handbook
   Synchro & Resolver Conversion

ADSP-2100 FAMILY
USER'S MANUAL

P T R PRENTICE HALL, Englewood Cliffs, NJ 07632

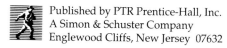 Published by PTR Prentice-Hall, Inc.
A Simon & Schuster Company
Englewood Cliffs, New Jersey 07632

The publisher offers discounts on this book when ordered
in bulk quantities. For more information, contact:

> Corporate Sales Department
> PTR Prentice Hall
> 113 Sylvan Avenue
> Englewood Cliffs, NJ 07632
>
> Phone: 201-592-2863
> Fax: 201-592-2249

Printed in the United States of America

10   9   8   7   6   5   4   3   2   1

ISBN   0-13-006958-2

Prentice-Hall International (UK) Limited, *London*
Prentice-Hall of Australia Pty. Limited, *Sydney*
Prentice-Hall Canada Inc., *Toronto*
Prentice-Hall Hispanoamericana, S.A., *Mexico*
Prentice-Hall of India Private Limited, *New Delhi*
Prentice-Hall of Japan, Inc., *Tokyo*
Simon & Schuster Asia Pte. Ltd., *Singapore*
Editora Prentice-Hall do Brasil, Ltda., *Rio de Janeiro*

Contents ■

Contents

Contents

Contents

Contents

Contents

Contents

CHAPTER 11   PROGRAMMING MODEL

Contents

CHAPTER 12   INSTRUCTION SET REFERENCE

Contents

Contents

Contents

Contents

Preface

This book is intended to serve primarily as a guide and reference for the design engineer. As the title suggests, it is a technical manual which will find its home both on the office bookshelf and laboratory workbench. For the non-technical reader, the book can also provide an overall understanding of what DSP is all about. Other professionals who will find this text informative include industry analysts, marketing engineers, and general technologists.

The topics addressed include, but are not limited to, architectural features of the DSP processor family that facilitate signal processing and ease the task of programming. The primary message delivered in the manual's introduction is that digital signal processing is gaining acceptance as a mainstream technology. The first three chapters, detailing computation units, program controller, and data transfers, are the core of this book. These are derived from previous editions of the *ADSP-2100*, *ADSP-2101*, and *ADSP-2111 User's Manuals*. Later portions of the book discuss system and memory interfacing considerations and provide a programmer's model.

To serve as a technical reference, the *ADSP-2100 Family User's Manual* includes a complete instruction set reference and appendix of pin descriptions. Mathematical considerations are covered in the appendices for division exceptions and numeric formats as well as the computation units chapter. Finally, the hexadecimal bit patterns of processor control registers, so difficult to remember for any digital designer, are "spelled out" in a highly visual way at the very back of the book where she or he can quickly find them.

The ADSP-2100 Family consists of five processors at the time of publication of this manual: **ADSP-2100A**, **ADSP-2101**, **ADSP-2105**, **ADSP-2111**, and **ADSP-21msp50**. Note that the ADSP-2100A offers a slightly different architecture than the other family processors, all of which contain on-chip memory. As new variants are added to the product line, this manual can be used as a reference for the new devices. For current information and product literature, contact Analog Devices DSP Marketing at (617) 461-3881.

Acknowledgements

The substance of this text was provided by many, if not all, of the individuals of Analog Devices' DSP Group of the Systems IC Products Division. Primary contributors included members of Technical Publications, Applications Engineering, and Design Engineering. Adele Hastings provided virtually all drawings and layout as well as editing for all chapters of this book.

Special thanks also to Amy Mar, Jim McQuaid, Dan Sheingold, and Karen Gettman.

Norwood, Massachusetts   *Hans Rempel*

Introduction ■ 1

1.1 OVERVIEW

The ADSP-2100 family is a collection of programmable single-chip microprocessors and microcomputers which share a base architecture optimized for digital signal processing (DSP) and other high-speed numeric processing applications. The various family processors differ principally in the number and nature of on-chip peripherals they incorporate in addition to the base architecture. On-chip memory, a timer, serial port(s) and a parallel host interface port are available in various members of the family. In addition, one branch of the family includes an on-chip analog interface for mixed analog/digital signal processing applications.

This manual provides the information necessary to understand and evaluate the family's base architecture, and to determine which device from the family best meets your needs for a particular application. Together with the data sheets describing the individual devices, this manual provides all the information required to design a hardware system using the ADSP-2100 family. Complete reference material for programmers is also included.

1.1.1 Functional Units

Table 1.1, which can be found on the following page, lists the main functional units in the ADSP-2100 architecture, and shows which units are included in each of the currently available devices.

- *Computational Units*—Every device in the ADSP-2100 family contains three independent, full-function computational units: an arithmetic/logic unit (ALU), a multiplier/accumulator (MAC) and a barrel shifter. The computational units process 16-bit data directly and provide hardware support for multiprecision computation as well.

- *Data Address Generators and Program Sequencer*—Two dedicated address generators and a program sequencer supply addresses for on-chip or external memory access. The sequencer supports single-cycle conditional branching and executes program loops with zero overhead. Dual data address generators allow the processor to

1

1 Introduction

	2100	2101	2102	2105	2106	2111	21msp50	21msp51
Arithmetic/Logic Unit	✓	✓	✓	✓	✓	✓	✓	✓
Multiply/Accumulator	✓	✓	✓	✓	✓	✓	✓	✓
Barrel Shifter	✓	✓	✓	✓	✓	✓	✓	✓
2 Data Address Generators	✓	✓	✓	✓	✓	✓	✓	✓
Program Sequencer	✓	✓	✓	✓	✓	✓	✓	✓
Data Memory (words)		1K*	1K*	.5K*	.5K*	1K*	1K*	1K*
Program Memory (words)		2K*	2K**	1K*	1K**	2K*	2K*	4K**
Timer		✓	✓	✓	✓	✓	✓	✓
Serial Port 0		✓	✓	✓	✓	✓	✓	✓
Serial Port 1		✓	✓			✓	✓	✓
Host Interface Port						✓	✓	✓
Analog Interface							✓	✓

* RAM
** RAM/ROM

Table 1.1 ADSP-2100 Family Functional Units

generate simultaneous addresses for dual operand fetches. Together the sequencer and data address generators keep the computational units busy, maximizing throughput.

- *Memory*—The ADSP-2100 family uses a modified Harvard architecture, in which data memory stores data, and program memory stores both instructions and data. Many ADSP-2100 family processors contain RAM or ROM, so that a portion of the program memory space and a portion of the data memory space reside on-chip. The speed of the on-chip memory allows the processor to fetch two operands (one from data memory and one from program memory) and an instruction (from program memory) in a single cycle.

The ADSP-2100, which does not have on-chip memory, has other features to improve data throughput, including an on-chip instruction cache which can store 16 words. When executing a program loop short enough to be wholly contained in the cache, the ADSP-2100 works like a three-bus system, fetching two operands (one from external data memory and one from external program memory) and an instruction (from the cache) on the same cycle. Many algorithms are readily coded in loops of 16 instructions or less because of the high-level syntax of the ADSP-2100 family assembly language.

Introduction 1

- *Serial Ports*—One or two serial ports (SPORTs) provide a complete serial interface with hardware companding (data compression and expansion). Both μ-law and A-law companding are supported. The ports interface easily and directly to a wide variety of popular serial devices. Each port can generate an internal programmable clock or accept an external clock.

- *Timer*—A programmable timer/counter with 8-bit prescaler provides periodic interrupt generation.

- *Host Interface Port*—The Host Interface Port (HIP) allows direct connection (with no glue logic) to a host processor. The HIP is made up of 16 data pins and 11 control pins. The HIP is extremely flexible and has provisions to allow simple interface to a variety of host processors. For example, the Motorola 68000, the Intel 8051, or another ADSP-2100 family processor can be easily connected to the HIP.

- *Analog Interface*—Certain members of the ADSP-2100 family include on-chip support for mixed analog/digital signal processing. This support includes an analog-to-digital converter (ADC), a digital-to-analog converter (DAC), analog and digital filters, and a parallel interface with the processor base architecture. The converters use sigma-delta technology to capture wanted samples from a greatly oversampled signal.

The ADSP-2100 family architecture exhibits a high degree of parallelism, tailored to DSP requirements. In a single cycle, any device in the family can:

- Generate the next program address.
- Fetch the next instruction.
- Perform one or two data moves.
- Update one or two data address pointers.
- Perform a computation.

In that same cycle, processors which have the relevant functional units can also:

- Receive and/or transmit data via the serial port(s).
- Receive and/or transmit data via the host interface port.
- Receive and/or transmit data via the analog interface.

1 Introduction

1.1.2 Memory And System Interface

In each ADSP-2100 family device, four on-chip buses connect memory with the other functional units: Data Memory Address bus, Data Memory Data bus, Program Memory Address bus, and Program Memory Data bus. In the ADSP-2100, which does not have internal memory, all four of these buses are extended off-chip. In the devices which have internal memory, a single external memory address bus and and a single external data bus are extended off-chip; these buses can be used for either program or data memory access.

External devices can gain control of buses with bus request/grant signals (BR and BG). ADSP-2100 family processors with internal program memory can continue running while the buses are granted to another master, as long as an external memory operation is not required.

Family devices other than the ADSP-2100 support memory-mapped peripherals with programmable wait state generation.

Boot circuitry provides for loading on-chip program memory automatically after reset. This can be done either through the memory interface from a single low-cost EPROM, or through the host interface port from a host processor. Multiple programs can be selected and loaded with no additional hardware.

ADSP-2100 family processors differ in their response to user interrupts. In all cases, however, the program sequencer allows the processor to respond with minimum latency. Interrupts can be nested with no additional latency. External interrupts can be configured as edge- or level-sensitive. Internal interrupts can be generated from the timer, the host interface port, the serial ports, and the analog interface.

1.1.3 Instruction Set

With very few exceptions, the ADSP-2100 family shares a single unified instruction set designed for upward compatibility with higher-integration devices. Chapter 12, "Instruction Set Reference" details the instruction set.

The ADSP-2100 family instruction set provides flexible data moves. Multifunction instructions combine one or more data moves with a computation. Every instruction can be executed in a single processor cycle. The assembly language uses an algebraic syntax for readability and ease of coding. A comprehensive set of software and hardware tools supports program development.

4

Introduction   1

1.1.4   DSP Performance

Signal processing applications make special performance demands which distinguish DSP architectures from other microprocessor and microcontroller architectures. Not only must instruction execution be fast, but DSPs must also perform well in each of the following areas:

- *Fast and Flexible Arithmetic*—The ADSP-2100 family base architecture provides single-cycle computation for multiplication, multiplication with accumulation, arbitrary amounts of shifting, and standard arithmetic and logical operations. In addition, the arithmetic units allow for any sequence of computation so that a given DSP algorithm can be executed without being reformulated.

- *Extended Dynamic Range*—Extended sums-of-products, common in DSP algorithms, are supported in the multiply/accumulate units of the ADSP-2100 family. A 40-bit accumulator provides eight bits of protection against overflow in successive additions to ensure that no loss of data or range occurs; 256 overflows would have to occur before any data is lost. Special instructions are provided for implementing block floating-point scaling of data.

- *Single-Cycle Fetch of Two Operands*—In extended sums-of-products calculations, two operands are needed on each cycle to feed the calculation. All members of the ADSP-2100 family are able to sustain two-operand data throughput, whether the data is stored on-chip or off.

- *Hardware Circular Buffers*—A large class of DSP algorithms, including filters, requires circular buffers. The ADSP-2100 family base architecture includes hardware to handle address pointer wraparound, simplifying the implementation of circular buffers both on- and off-chip, and reducing overhead (thereby improving performance).

- *Zero-Overhead Looping and Branching*—DSP algorithms are repetitive and most logically expressed as loops. The program sequencer in the ADSP-2100 family supports looped code with zero overhead, combining excellent performance with the clearest program structure. Likewise, there are no overhead penalties for conditional branches.

1 Introduction

1.2 BASE ARCHITECTURE

This section describes the base architecture of the ADSP-2100 family, as shown in Figure 1.1.

Each component in the base architecture is described in detail in Part 2 of this book, as shown in the chart below:

Component	Chapter
Arithmetic/logic unit	2
Multiplier/accumulator	2
Barrel shifter	2
Program sequencer	3
Status registers and stacks	3
Two data address generators	4
PMD-DMD bus exchange	4

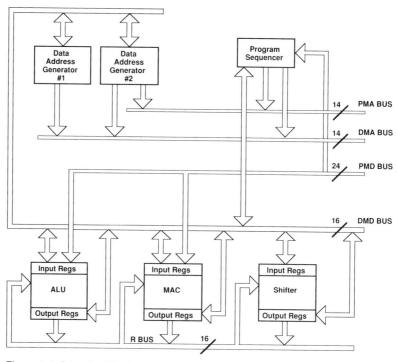

Figure 1.1 Base Architecture

Introduction 1

1.2.1 Computational Units

Every device in the ADSP-2100 family contains three independent, full-function computational units: an arithmetic/logic unit (ALU), a multiplier/accumulator (MAC) and a barrel shifter. The computation units process 16-bit data directly and provide hardware support for multiprecision computation as well.

The ALU performs a standard set of arithmetic and logic operations in addition to division primitives. The MAC performs single-cycle multiply, multiply/add and multiply/subtract operations. The shifter performs logical and arithmetic shifts, normalization, denormalization, and derive-exponent operations. The shifter implements numeric format control including multiword floating-point representations. The computational units are arranged side-by-side instead of serially so that the output of any unit may be the input of any unit on the next cycle. The internal result (R) bus directly connects the computational units to make this possible.

All three units contain input and output registers which are accessible from the internal data memory data (DMD) bus. Computational operations generally take their operands from input registers and load the result into an output register. The registers act as a stopover point for data between memory and the computational circuitry. This feature introduces one level of pipelining on input, and one level on output. The R bus allows the result of a previous computation to be used directly as the input to another computation. This avoids excessive pipeline delays when a series of different operations are performed.

1.2.2 Address Generators And Program Sequencer

Two dedicated data address generators and a powerful program sequencer ensure efficient use of the computational units. The data address generators (DAGs) provide memory addresses when memory data is transferred to or from the input or output registers. Each DAG keeps track of up to four address pointers. When a pointer is used for indirect addressing, it is post-modified by a value in a specified register. With two independent DAGs, the processor can generate two addresses simultaneously for dual operand fetches.

A length value may be associated with each pointer to implement automatic modulo addressing for circular buffers. (The circular buffer feature is also used by the serial ports and the analog interface for automatic data transfers. Refer to the chapter on serial ports for additional information.)

1 Introduction

DAG1 can supply addresses to data memory only; DAG2 can supply addresses to either the data memory or the program memory. When the appropriate mode bit is set in the mode status register (MSTAT), the output address of DAG1 is bit-reversed before being driven onto the address bus. This feature facilitates addressing in radix-2 Fast Fourier Transform (FFT) algorithms.

The program sequencer supplies instruction addresses to the program memory. The sequencer is driven by the instruction register which holds the currently executing instruction. The instruction register introduces a single level of pipelining into the program flow. Instructions are fetched and loaded into the instruction register during one processor cycle, and executed during the following cycle while the next instruction is prefetched. To minimize overhead cycles, the sequencer supports conditional jumps, subroutine calls and returns in a single cycle. With an internal loop counter and loop stack, the processor executes looped code with zero overhead. No explicit jump instructions are required to loop.

1.2.3 Buses

The internal components are supported by five internal buses. The program memory address (PMA) and data memory address (DMA) buses are used internally for the addresses associated with program and data memory. The program memory data (PMD) and data memory data (DMD) buses are used for the data associated with the memory spaces. In the ADSP-2100, these four buses are all extended off-chip. In devices with internal memory, they are multiplexed into a single external address bus and a single external data bus; the $\overline{BMS}$, $\overline{DMS}$ and $\overline{PMS}$ signals select the different address spaces. The R bus transfers intermediate results directly between the various computational units.

The PMA bus is 14 bits wide allowing direct access of up to 16K words of mixed instruction code and data. The PMD bus is 24 bits wide to accommodate the 24-bit instruction width.

The DMA bus is 14 bits wide allowing direct access of up to 16 K words of data. The data memory data (DMD) bus is 16 bits wide. The DMD bus provides a path for the contents of any register in the processor to be transferred to any other register or to any data memory location in a single cycle. The data memory address comes from two sources: an absolute value specified in the instruction code (direct addressing) or the output of a data address generator (indirect addressing). Only indirect addressing is supported for data fetches from program memory.

Introduction   1

The program memory data (PMD) bus can also be used to transfer data to and from the computational units through direct paths or via the PMD-DMD bus exchange unit. The PMD-DMD bus exchange unit permits data to be passed from one bus to the other. It contains hardware to overcome the 8-bit width discrepancy between the two buses, when necessary.

1.3   ON-CHIP PERIPHERALS

This section describes the additional functional units which are included in some, but not all, members of the ADSP-2100 family.

1.3.1   Serial Ports

Most family processors have one or two bidirectional, double-buffered serial ports (SPORTs) for serial communications. The SPORTs are synchronous and use framing signals to control data flow. Each SPORT can generate its serial clock internally or use an external clock. The framing sync signals may be generated internally or by an external device. Word lengths may vary from three to sixteen bits. One SPORT (SPORT0) has a multichannel capability which allows the receiving or transmitting of arbitrary data words from a 24-word or 32-word bitstream. In those devices with two SPORTs, the second (SPORT1) may optionally be configured as two additional external interrupt pins ($\overline{IRQ1}$ and $\overline{IRQ0}$)and the Flag Out (FO) and Flag In (FI) pins.

1.3.2   Timer

The programmable interval timer provides periodic interrupt generation. An 8-bit prescaler register allows the timer to decrement a 16-bit count register over a range from each cycle to every 256 cycles. An interrupt is generated when this count register reaches zero. The count register is automatically reloaded from a 16-bit period register and the count resumes immediately.

1.3.3   Host Interface Port

The host interface port (HIP) is a parallel I/O port that allows for an easy connection to a host processor. Through the HIP, an ADSP-2100 family processor can be used as a memory-mapped peripheral to a host computer. The HIP operates in parallel with and asynchronous to the ADSP-2100 family processing base architecture. The host interface port consists of registers through which the ADSP-2100 family processor and the host computer pass data and status information. The HIP can be configured for: an 8-bit data bus or 16-bit data bus; a multiplexed address/data bus or separate address and data buses; and separate read and write strobes or a read/write strobe and a data strobe.

1   Introduction

1.3.4   Analog Interface

The analog interface consists of input amplifiers and a 16-bit sigma-delta analog-to-digital converter (ADC) as well as a sigma-delta digital-to-analog converter (DAC) and a differential output amplifier. A set of memory-mapped registers is used to control the operation of the analog section and to pass data between the analog section and the processor base architecture.

1.4   ADSP-2100 FAMILY DEVELOPMENT SYSTEM

The ADSP-2100 family is supported with a complete set of software and hardware development tools. The ADSP-2100 Family Development System includes the Development Software for software design and an Emulator and EZ-Tools™ for hardware debugging.

The Development Software includes:

- *System Builder*—The System Builder defines the architecture of systems under development. This includes the specification of the amount of external RAM/ROM memory available and any memory-mapped I/O ports for the target hardware environment as well as the allocation of program and data memory.

- *Assembler*—The Assembler assembles the source code and data modules as well as supporting the high-level syntax of the instruction set. In addition to supporting a full range of system diagnostics, the Assembler provides flexible macro processing, include files, and modular code development.

- *Linker*—The Linker links separately assembled modules. It maps the linked code and data output to the target system hardware, as specified by the System Builder output.

- *Simulator*—The Simulator performs an interactive, instruction-level simulation of the hardware configuration described by the System Builder. It flags illegal operations and supports full symbolic assembly and disassembly.

Introduction   1

- *PROM Splitter*—This module reads the Linker output and generates PROM programmer compatible files.

- *C Compiler*—The C Compiler reads ANSI C source and outputs ADSP-2100 family source code ready to be assembled. It also supports inline assembler code.

Emulators and EZ-Tools provide hardware debugging of ADSP-2100 family systems. Emulators perform stand-alone in-circuit emulation, using an ADSP-2100 family device in self-emulation mode. The Emulator design provides execution with little or no degradation in processor performance. EZ-Tools are low-cost, basic systems. EZ-ICE™ is a simplified emulator that provides a subset of emulation functions. EZ-LAB™ is an evaluation board for running example applications.

For additional information on the Development System, refer to the *ADSP-2100 Family Assembler Manual*, *ADSP-2100 Family Simulator Manual*, and the *ADSP-2100 Family C Compiler Manual*.

1.5   ORGANIZATION OF THIS MANUAL
The rest of this manual is organized as follows.

Part 2, "Base Architecture," describes the principal architectural features shared by all members of the ADSP-2100 family. Part 2 consists of three chapters:

- Chapter 2, "Computational Units," describes the functions and internal organization of the arithmetic/logic unit (ALU), the multiplier/accumulator (MAC), and the barrel shifter.

- Chapter 3, "Program Control," describes the program sequencer, interrupt controller and status and condition logic.

- Chapter 4, "Data Transfer," describes the data address generators (DAGs) and the PMD-DMD bus exchange unit.

1 Introduction

Part 3, "On-Chip Peripherals," describes the additional functional units which are included in some, but not all, members of the ADSP-2100 family. (See Table 1.1 for a list of the units included in each device.) Part 3 consists of four chapters:

- Chapter 5, "Serial Ports," describes the serial ports, SPORT0 and SPORT1.

- Chapter 6, "Timer," explains the programmable interval timer.

- Chapter 7, "Host Interface Port," describes the operation of the host interface port, including (optional) boot loading and software reset.

- Chapter 8, "Analog Interface," describes the operation and the internal architecture of the analog interface.

Part 4, "Hardware Interface," describes the behavior of the ADSP-2100 family processors from the point of view of external memory and control logic. Part 4 consists of two chapters:

- Chapter 9, "System Interface," discusses the issue of system clocking, and describes the processors' control interface, the software reboot function, and the powerdown mode. (The powerdown mode is available in the mixed-signal processors.)

- Chapter 10, "Memory Interface," describes the data and program memory spaces. For family processors other than the ADSP-2100, this chapter describes both internal and external memory, including the use of boot memory space. A special section is devoted to the ADSP-2100, since its memory interface differs from that of the other family processors. For timing characteristics, refer to the data sheets for the individual devices.

Part 5, "Programmer's Reference," describes the ADSP-2100 family from the point of view of software. Additional programmer's reference material is in the *ADSP-2100 Family Assembler Manual*. Part 5 consists of two chapters:

- Chapter 11, "Programming Model," gives a functional description of the processor resources—such as registers—as they appear to software.

- Chapter 12, "Instruction Set Reference," provides a detailed description of each ADSP-2100 family instruction.

Introduction   1

Chapter 13, "Hardware Examples," gives examples of system design using the ADSP-2100 family. Each example illustrates the solution to a different system-design problem, using block diagrams, explanatory text, and programs or timing diagrams as needed.

Chapter 14, "Software Examples," provides illustrative ADSP-2100 family code for some important DSP and other numerical algorithms.

The Appendices provide reference material and further details on specific issues:

- Appendix A, "Instruction Coding," gives the complete set of opcodes and specifies the bit patterns for choices within each field of the instruction word.

- Appendix B, "Division Exceptions," describes signed and unsigned division.

- Appendix C, "Numeric Formats," describes the fixed-point numerical formats directly supported by the ADSP-2100 family, discusses block floating-point arithmetic, and tells how to handle the results of multiplication for operands of various formats.

- Appendix D, "Pin Descriptions," describes the pin functions and pin configurations of the various family processors.

- Appendix E, "Control/Status Registers," summarizes the contents and locations of all control and status registers.

Computational Units ◨ 2

2.1 OVERVIEW

This chapter describes the architecture and function of the three computational units: the arithmetic/logic unit, the multiplier/accumulator and the barrel shifter.

Every device in the ADSP-2100 family is a 16-bit, fixed-point machine. Most operations assume a twos-complement number representation, while others assume unsigned numbers or simple binary strings. Special features support multiword arithmetic and block floating-point. Details concerning the various number formats supported by the ADSP-2100 family are given in Appendix C.

In ADSP-2100 family arithmetic, signed numbers are always in twos-complement format. The family does not use signed-magnitude, ones-complement, BCD or excess-n formats.

2.1.1 Binary String

This is the simplest binary notation; sixteen bits are treated as a bit pattern. Examples of computation using this format are the logical operations: NOT, AND, OR, XOR. These ALU operations treat their operands as binary strings with no provision for sign bit or binary point placement.

2.1.2 Unsigned

Unsigned binary numbers may be thought of as positive, having nearly twice the magnitude of a signed number of the same length. The least significant words of multiple precision numbers are treated as unsigned numbers.

2.1.3 Signed Numbers: Twos-Complement

In discussions of ADSP-2100 family arithmetic, "signed" refers to twos-complement. Most ADSP-2100 family operations presume or support twos-complement arithmetic. The ADSP-2100 family does not use signed-magnitude, ones-complement, BCD or excess-n formats.

15

2 Computational Units

2.1.4   Fractional Representation: 1.15

ADSP-2100 family arithmetic is optimized for numerical values in a fractional binary format denoted by 1.15 ("one dot fifteen"). In the 1.15 format, there is one sign bit (the MSB) and fifteen fractional bits representing values from –1 up to one LSB less than +1.

Figure 2.1 shows the bit weighting for 1.15 numbers. Below are examples of 1.15 numbers and their decimal equivalents.

1.15 Number	Decimal Equivalent
0x0001	0.000031
0x7FFF	0.999969
0xFFFF	–0.000031
0x8000	–1.000000

-2^0	2^{-1}	2^{-2}	2^{-3}	2^{-4}	2^{-5}	2^{-6}	2^{-7}	2^{-8}	2^{-9}	2^{-10}	2^{-11}	2^{-12}	2^{-13}	2^{-14}	2^{-15}

Figure 2.1 Bit Weighting For 1.15 Numbers

2.1.5   ALU Arithmetic

All operations on the ALU treat operands and results as simple 16-bit binary strings, except the signed division primitive (DIVS). Various status bits treat the results as signed: the overflow (AV) condition code, and the negative (AN) flag.

The logic of the overflow bit (AV) is based on twos-complement arithmetic. It is set if the MSB changes in a manner not predicted by the signs of the operands and the nature of the operation. For example, adding two positive numbers must generate a positive result; a change in the sign bit signifies an overflow and sets AV. Adding a negative and a positive may result in either a negative or positive result, but cannot overflow.

The logic of the carry bit (AC) is based on unsigned-magnitude arithmetic. It is set if a carry is generated from bit 16 (the MSB). The (AC) bit is most useful for the lower word portions of a multiword operation.

Computational Units 2

2.1.6 MAC Arithmetic

The multiplier produces results that are binary strings. The inputs are "interpreted" according to the information given in the instruction itself (signed times signed, unsigned times unsigned, a mixture, or a rounding operation). The 32-bit result from the multiplier is assumed to be signed, in that it is sign-extended across the full 40-bit width of the MR register set.

Except for the ADSP-2100 processor, the ADSP-2100 family supports two modes of format adjustment: the fractional mode for fractional operands, 1.15 format (1 signed bit, 15 fractional bits), and the integer mode for integer operands, 16.0 format.

When the processor multiplies two 1.15 operands, the result is a 2.30 (2 sign bits, 30 fractional bits) number. In the fractional mode, the MAC automatically shifts the multiplier product (P) left one bit before transferring the result to the multiplier result register (MR). This shift causes the multiplier result to be in 1.31 format, which can be rounded to 1.15 format. Figure 2.7, in the MAC section of this chapter, shows this.

In the integer mode, the left shift does not occur. For example, if the operands are in the 16.0 format, the 32-bit multiplier result would be in 32.0 format. A left shift is not needed; it would change the numerical representation. Figure 2.8 in the MAC section of this chapter shows this.

The ADSP-2100 does not support the integer mode, so format adjustment must be performed by software when two integers are multiplied. Typically, this means shifting the result down (right) one bit to get the correct, 32-bit twos-complement value. Since the MAC output register set stores 40 bits, this result is not lost and can be retrieved with the shifter.

2.1.7 Shifter Arithmetic

Many operations in the shifter are explicitly geared to signed (twos-complement) or unsigned values: logical shifts assume unsigned-magnitude or binary string values and arithmetic shifts assume twos-complement.

The exponent logic assumes twos-complement numbers. The exponent logic supports block floating-point, which is also based on twos-complement fractions.

2 Computational Units

2.1.8 Summary

Table 2.1 summarizes some of the arithmetic characteristics of ADSP-2100 family operations. In addition to the numeric types described in this section, the ADSP-2100 Family C Compiler supports a form of 32-bit floating-point in which one 16-bit word is the exponent and the other word is the mantissa. See the *ADSP-2100 Family C Compiler Manual*.

OPERATION	ARITHMETIC FORMATS	
ALU	*Operands*	*Result*
Addition	Signed or unsigned	Interpret flags
Subtraction	Signed or unsigned	Interpret flags
Logical Operations	Binary string	same as operands
Division	Explicitly signed/unsigned	same as operands
ALU Overflow	Signed	same as operands
ALU Carry Bit	16-bit unsigned	same as operands
ALU Saturation	Signed	same as operands
MAC, Fractional		
Multiplication (P)	1.15 Explicitly signed/unsigned	32 bits (2.30)
Multiplication (MR)	1.15 Explicitly signed/unsigned	2.30 shifted to 1.31
Mult / Add	1.15 Explicitly signed/unsigned	2.30 shifted to 1.31
Mult / Subtract	1.15 Explicitly signed/unsigned	2.30 shifted to 1.31
MAC Saturation	Signed	same as operands
MAC, Integer Mode [Not available in ADSP-2100]		
Multiplication (P)	1.15 Explicitly signed/unsigned	32 bits (2.30)
Multiplication (MR)	16.0 Explicitly signed/unsigned	32.0 no shift
Mult / Add	16.0 Explicitly signed/unsigned	32.0 no shift
Mult / Subtract	16.0 Explicitly signed/unsigned	32.0 no shift
MAC Saturation	Signed	same as operands
Shifter		
Logical Shift	Unsigned / binary string	same as operands
Arithmetic Shift	Signed	same as operands
Exponent Detection	Signed	same as operands

Table 2.1   Arithmetic Formats

Computational Units   2

2.2   ARITHMETIC/LOGIC UNIT (ALU)

The arithmetic/logic unit (ALU) provides a standard set of arithmetic and logical functions. The arithmetic functions are add, subtract, negate, increment, decrement and absolute value. These are supplemented by two division primitives with which multiple cycle division can be constructed. The logic functions are AND, OR, XOR (exclusive OR) and NOT.

2.2.1   ALU Block Diagram Discussion

Figure 2.2, on the following page, shows a block diagram of the ALU.

The ALU is 16 bits wide with two 16-bit input ports, X and Y, and one output port, R. The ALU accepts a carry-in signal (CI) which is the carry bit from the processor arithmetic status register (ASTAT). The ALU generates six status signals: the zero (AZ) status, the negative (AN) status, the carry (AC) status, the overflow (AV) status, the X-input sign (AS) status, and the quotient (AQ) status. All arithmetic status signals are latched into the arithmetic status register (ASTAT) at the end of the cycle. Please see the "Instruction Set Reference" chapter of the *ADSP-2100 Family Assembler Manual* for information on how each instruction affects the ALU flags.

The X input port of the ALU can accept data from two sources: the AX register file or the result (R) bus. The R bus connects the output registers of all the computational units, permitting them to be used as input operands directly. The AX register file is dedicated to the X input port and consists of two registers, AX0 and AX1. These AX registers are readable and writable from the DMD bus. The instruction set also provides for reading these registers over the PMD bus, but there is no direct connection; this operation uses the DMD-PMD bus exchange unit. The AX register file outputs are dual-ported so that one register can provide input to the ALU while either one simultaneously drives the DMD bus.

The Y input port of the ALU can also accept data from two sources: the AY register file and the ALU feedback (AF) register. The AY register file is dedicated to the Y input port and consists of two registers, AY0 and AY1. These registers are readable and writable from the DMD bus and writable from the PMD bus. The instruction set also provides for reading these registers over the PMD bus, but there is no direct connection; this operation uses the DMD-PMD bus exchange unit. The AY register file outputs are also dual-ported: one AY register can provide input to the ALU while either one simultaneously drives the DMD bus.

The output of the ALU is loaded into either the ALU feedback (AF) register or the ALU result (AR) register. The AF register is an ALU

2 Computational Units

internal register which allows the ALU result to be used directly as the ALU Y input. The AR register can drive both the DMD bus and the R bus. It is also loadable directly from the DMD bus. The instruction set also provides for reading AR over the PMD bus, but there is no direct connection; this operation uses the DMD-PMD bus exchange unit.

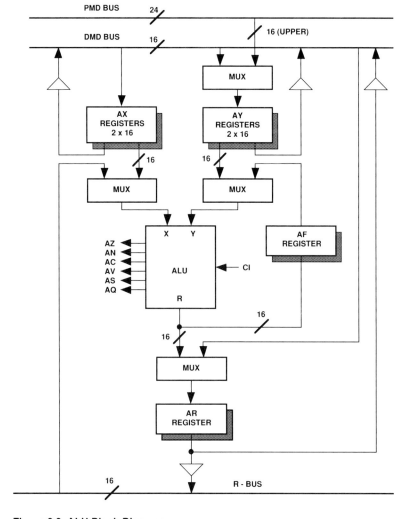

Figure 2.2 ALU Block Diagram

Computational Units   2

Any of the registers associated with the ALU can be both read and written in the same cycle. Registers are read at the beginning of the cycle and written at the end of the cycle. A register read, therefore, reads the value loaded at the end of a previous cycle. A new value written to a register cannot be read out until a subsequent cycle. This allows an input register to provide an operand to the ALU at the beginning of the cycle and be updated with the next operand from memory at the end of the same cycle. It also allows a result register to be stored in memory and updated with a new result in the same cycle. See the discussion of "Multifunction Instructions" in Chapter 12, "Instruction Set Reference" for an illustration of this same-cycle read and write.

The ALU section contains a duplicate bank of registers, shown in Figure 2.2 behind the primary registers. There are actually two sets of AR, AF, AX, and AY register files. Only one bank is accessible at a time. The additional bank of registers can be activated (such as during an interrupt service routine) for extremely fast context switching. A new task, like an interrupt service routine, can be executed without transferring current states to storage.

The selection of the primary or alternate bank of registers is controlled by bit 0 in the processor mode status register (MSTAT). If this bit is a 0, the primary bank is selected; if it is a 1, the secondary bank is selected.

2.2.2   Standard Functions
The standard ALU functions are listed below.

$R = X + Y$	Add X and Y operands
$R = X + Y + CI$	Add X and Y operands and carry-in bit
$R = X - Y$	Subtract Y from X operand
$R = X - Y + CI - 1$	Subtract Y from X operand with "borrow"
$R = Y - X$	Subtract X from Y operand
$R = Y - X + CI - 1$	Subtract X from Y operand with "borrow"
$R = -X$	Negate X operand *(twos-complement)*
$R = -Y$	Negate Y operand *(twos-complement)*
$R = Y + 1$	Increment Y operand
$R = Y - 1$	Decrement Y operand
$R = PASS X$	Pass X operand to result unchanged
$R = PASS Y$	Pass Y operand to result unchanged
$R = 0$ *(PASS 0)*	Clear result to zero
$R = ABS X$	Absolute value of X operand
$R = X AND Y$	Logical AND of X and Y operands

2 Computational Units

R = X OR Y	Logical OR of X and Y operands
R = X XOR Y	Logical Exclusive OR of X and Y operands
R = NOT X	Logical NOT of X operand *(ones-complement)*
R = NOT Y	Logical NOT of Y operand *(ones-complement)*

2.2.3   ALU Input/Output Registers

The sources of ALU input and output registers are shown below.

Source for X input port	*Source for Y input port*	*Destination for R output port*
AX0, AX1	AY0, AY1	AR
AR	AF	AF
MR0, MR1, MR2		
SR0, SR1		

MR0, MR1 and MR2 are multiplier/accumulator result registers; SR0 and SR1 are shifter result registers.

2.2.4   Multiprecision Capability

Multiprecision operations are supported in the ALU with the carry-in signal and ALU carry (AC) status bit. The carry-in signal is the AC status bit that was generated by a previous ALU operation. The "add with carry" (+ C) operation is intended for adding the upper portions of multiprecision numbers. The "subtract with borrow" (C – 1 is effectively a "borrow") operation is intended for subtracting the upper portions of multiprecision numbers.

2.2.5   ALU Saturation Mode

The AR register has a twos-complement saturation mode of operation which automatically sets it to the maximum negative or positive value if an ALU result overflows or underflows. This feature is enabled by setting bit 3 of the mode status register (MSTAT). When enabled, the value loaded into AR during an ALU operation depends on the state of the overflow and carry status generated by the ALU on that cycle. The following table summarizes the loading of the AR when the saturation mode is enabled.

Overflow (AV)	*Carry (AC)*	*AR Contents*	
0	0	ALU Output	
0	1	ALU Output	
1	0	0111111111111111	*full-scale positive*
1	1	1000000000000000	*full-scale negative*

Table 2.2 Saturation Mode

Computational Units   2

The operation of the ALU saturation mode is in contrast to the Multiplier/ Accumulator saturation ability, which is enabled only on an instruction by instruction basis. For the ALU, enabling saturation means that all subsequent operations are processed this way.

When the ALU saturation mode is used, only the AR register saturates; if the AF register is the destination, wrap-around will occur but the flags will reflect the saturated result.

2.2.6   ALU Overflow Latch Mode

The ALU overflow latch mode, enabled by setting bit 2 in the mode status register (MSTAT), causes the AV bit to "stick" once it is set. In this mode, when an ALU overflow occurs, AV will be set and remain set, even if subsequent ALU operations do not generate overflows. In this mode, AV can only be cleared by writing a zero to it directly from the DMD bus.

2.2.7   Division

The ALU supports division. The divide function is achieved with additional shift circuitry not shown in Figure 2.2, the block diagram. Division is accomplished with two special divide primitives. These are used to implement a non-restoring conditional add-subtract division algorithm. The division can be either signed or unsigned; however, the dividend and divisor must both be of the same type. Appendix B details various exceptions to the normal division operation as described in this section.

A single-precision divide, with a 32-bit dividend (numerator) and a 16-bit divisor (denominator), yielding a 16-bit quotient, executes in 16 cycles. Higher and lower precision quotients can also be calculated. The divisor can be stored in AX0, AX1 or any of the R registers. The upper half of a signed dividend can start in either AY1 or AF. The upper half of an unsigned dividend must be in AF. The lower half of any dividend must be in AY0. At the end of the divide operation, the quotient will be in AY0.

The first of the two primitive instructions "divide-sign" (DIVS) is executed at the beginning of the division when dividing signed numbers. This operation computes the sign bit of the quotient by performing an exclusive-OR of the sign bits of the divisor and the dividend. The AY0 register is shifted one place so that the computed sign bit is moved into the LSB position. The computed sign bit is also loaded into the AQ bit of the arithmetic status register. The MSB of AY0 shifts into the LSB position of AF, and the upper 15 bits of AF are loaded with the lower 15 R bits

2 Computational Units

from the ALU, which simply passes the Y input value straight through to the R output. The net effect is to left shift the AF-AY0 register pair and move the quotient sign bit into the LSB position. The operation of DIVS is illustrated in Figure 2.3.

When dividing unsigned numbers, the DIVS operation is not used. Instead, the AQ bit in the arithmetic status register (ASTAT) should be initialized to zero by manually clearing it. The AQ bit indicates to the following operations that the quotient should be assumed positive.

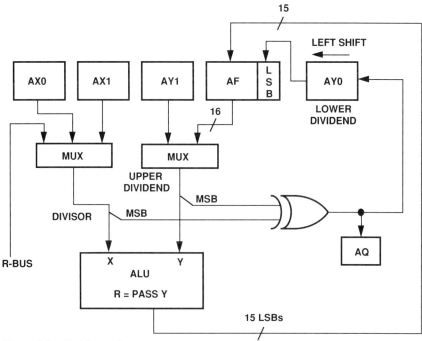

Figure 2.3 DIVS Operation

The second division primitive is the "divide-quotient" (DIVQ) instruction which generates one bit of quotient at a time and is executed repeatedly to compute the remaining quotient bits. For unsigned single precision divides, the DIVQ instruction is executed 16 times to produce 16 quotient bits. For signed single precision divides, the DIVQ instruction is executed 15 times after the sign bit is computed by the DIVS operation. DIVQ instruction shifts the AY0 register left by one bit so that the new quotient

Computational Units   2

bit can be moved into the LSB position. The status of the AQ bit generated
from the previous operation determines the ALU operation to calculate the
partial remainder. If AQ = 1, the ALU adds the divisor to the partial
remainder in AF. If AQ = 0, the ALU subtracts the divisor from the partial
remainder in AF. The ALU output R is offset loaded into AF just as with the
DIVS operation. The AQ bit is computed as the exclusive-OR of the divisor
MSB and the ALU output MSB, and the quotient bit is this value inverted.
The quotient bit is loaded into the LSB of the AY0 register which is also
shifted left by one bit. The DIVQ operation is illustrated in Figure 2.4.

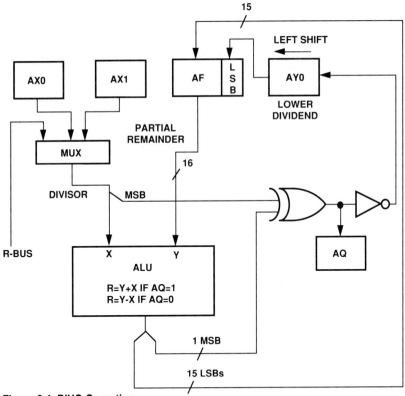

Figure 2.4 DIVQ Operation

2 Computational Units

The format of the quotient for any numeric representation can be determined by the format of the dividend and divisor. Let NL represent the number of bits to the left of the binary point, and NR represent the number of bits to the right of the binary point of the dividend; DL represent the number of bits to the left of the binary point, and DR represent the number of bits to the right of the binary point of the divisor; then the quotient has NL–DL+1 bits to the left of the binary point and NR–DR–1 bits to the right of the binary point.

Some format manipulation may be necessary to guarantee the validity of the quotient. For example, if both operands are signed and fully fractional (dividend in 1.31 format and divisor in 1.15 format) the result is fully fractional (in 1.15 format) and therefore the dividend must be smaller than the divisor for a valid result.

To divide two integers (dividend in 32.0 format and divisor in 16.0 format) and produce an integer quotient (in 16.0 format), you must shift the dividend one bit to the left (into 31.1 format) before dividing. Additional discussion and code examples can be found in *Digital Signal Processing Applications Using the ADSP-2100 Family*.

Dividend	BBBBB.BBBBBBBBBBBBBBBBBBBBBBBBBBB
	NL bits NR bits
Divisor	BB.BBBBBBBBBBBBBB
	DL bits DR bits
Quotient	BBBB.BBBBBBBBBBBB
	(NL-DL+1) bits (NR-DR-1) bits

Figure 2.5 Quotient Format

The algorithm overflows if the result cannot be represented in the format of the quotient as calculated above or when the divisor is zero or less than the dividend in magnitude.

Computational Units   2

2.2.8   ALU Status

The ALU status bits in the ASTAT register are defined below. Complete information about the ASTAT register and specific bit mnemonics and positions is provided in the Program Control chapter.

Flag	Name	Definition
AZ	Zero	Logical NOR of all the bits in the ALU result register. True if ALU output equals zero.
AN	Negative	Sign bit of the ALU result. True if the ALU output is negative.
AV	Overflow	Exclusive-OR of the carry outputs of the two most significant adder stages. True if the ALU overflows.
AC	Carry	Carry output from the most significant adder stage.
AS	Sign	Sign bit of the ALU X input port. Affected only by the ABS instruction.
AQ	Quotient	Quotient bit generated only by the DIVS and DIVQ instructions.

2.3   MULTIPLIER/ACCUMULATOR (MAC)

The multiplier/accumulator (MAC) provides high-speed multiplication, multiplication with cumulative addition, multiplication with cumulative subtraction, saturation and clear-to-zero functions. A feedback function allows part of the accumulator output to be directly used as one of the multiplicands on the next cycle.

2.3.1   MAC Block Diagram Discussion

Figure 2.6, which can be found on the following page, shows a block diagram of the multiplier/accumulator section.

The multiplier has two 16-bit input ports X and Y, and a 32-bit product output port P. The 32-bit product is passed to a 40-bit adder/subtracter which adds or subtracts the new product from the content of the multiplier result (MR) register, or passes the new product directly to MR. The MR register is 40 bits wide. In this manual, we refer to the entire register as MR. The register actually consists of three smaller registers: MR0 and MR1 which are 16 bits wide and MR2 which is 8 bits wide.

The adder/subtracter is greater than 32 bits to allow for intermediate overflow in a series of multiply/accumulate operations. The multiply overflow (MV) status bit is set when the accumulator has overflowed

2 Computational Units

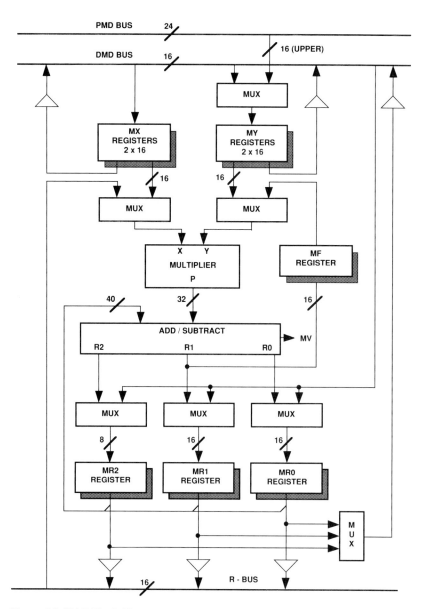

Figure 2.6 MAC Block Diagram

Computational Units   2

beyond the 32-bit boundary, that is, when there are significant (non-sign) bits in the top nine bits of the MR register (based on twos-complement arithmetic).

The input/output registers of the MAC are similar to the ALU.

The X input port can accept data from either the MX register file or from any register on the result (R) bus. The R bus connects the output registers of all the computational units, permitting them to be used as input operands directly. There are two registers in the MX register file, MX0 and MX1. These registers can be read and written from the DMD bus. The MX register file outputs are dual-ported so that one register can provide input to the multiplier while either one simultaneously drives the DMD bus.

The Y input port can accept data from either the MY register file or the MF register. The MY register file has two registers, MY0 and MY1; these registers can be read and written from the DMD bus and written from the PMD bus. The instruction set also provides for reading these registers over the PMD bus, but there is no direct connection; this operation uses the DMD-PMD bus exchange unit. The MY register file outputs are also dual-ported so that one register can provide input to the multiplier while either one simultaneously drives the DMD bus.

The output of the adder/subtracter goes to either the MF register or the MR register. The MF register is a feedback register which allows bits 16–31 of the result to be used directly as the multiplier Y input on a subsequent cycle. The 40-bit adder/subtracter register (MR) is divided into three sections: MR2, MR1, and MR0. Each of these registers can be loaded directly from the DMD bus and output to either the DMD bus or the R bus.

Any of the registers associated with the MAC can be both read and written in the same cycle. Registers are read at the beginning of the cycle and written at the end of the cycle. A register read, therefore, reads the value loaded at the end of a previous cycle. A new value written to a register cannot be read out until a subsequent cycle. This allows an input register to provide an operand to the MAC at the beginning of the cycle and be updated with the next operand from memory at the end of the same cycle. It also allows a result register to be stored in memory and updated with a new result in the same cycle. See the discussion of "Multifunction Instructions" in Chapter 12 "Instruction Set Reference" for an illustration of this same-cycle read and write.

2   Computational Units

The MAC section contains a duplicate bank of registers, shown in Figure 2.6 behind the primary registers. There are actually two sets of MR, MF, MX, and MY register files. Only one bank is accessible at a time. The additional bank of registers can be activated for extremely fast context switching. A new task, such as an interrupt service routine, can be executed without transferring current states to storage.

The selection of the primary or alternate bank of registers is controlled by bit 0 in the processor mode status register (MSTAT). If this bit is a 0, the primary bank is selected; if it is a 1, the secondary bank is selected.

2.3.2   MAC Operations

This section explains the functions of the MAC, its input formats and its handling of overflow and saturation.

2.3.2.1 Standard Functions

The functions performed by the MAC are:

X*Y   Multiply X and Y operands.
MR+X*Y   Multiply X and Y operands and add result to MR register.
MR-X*Y   Multiply X and Y operands and subtract result from MR register.
0   Clear result (MR) to zero.

Except for the ADSP-2100 processor, the ADSP-2100 family provides two modes for the standard multiply/accumulate function: fractional mode for fractional numbers (1.15), and integer mode for integers (16.0). The ADSP-2100 supports only the fractional mode.

In the fractional mode, the 32-bit P output is format adjusted, that is, sign-extended and shifted one bit to the left before being added to MR. For example, bit 31 of P lines up with bit 32 of MR (which is bit 0 of MR2) and bit 0 of P lines up with bit 1 of MR (which is bit 1 of MR0). The LSB is zero-filled. The fractional multiplier result format is shown in Figure 2.7.

Computational Units   2

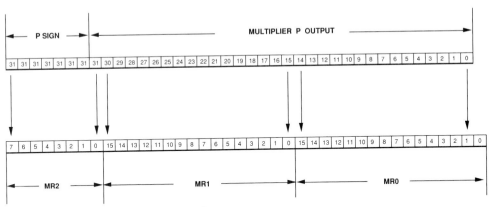

Figure 2.7 Fractional Multiplier Result Format

In the integer mode, the 32-bit P register is not shifted before being added to MR. Figure 2.8 shows the integer-mode result placement.

The mode is selected by bit 4 of the mode status register (MSTAT). In devices other than the ADSP-2100, if this bit is a 1, the integer mode is selected. Otherwise, the fractional mode is selected. In either mode, the multiplier output P is fed into a 40-bit adder/subtracter which adds or subtracts the new product with the current contents of the MR register to form the final 40-bit result R.

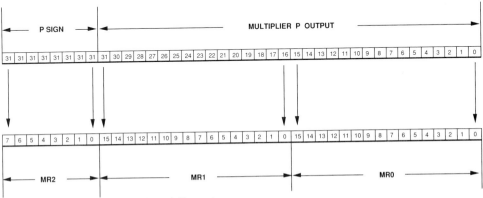

Figure 2.8 Integer Multiplier Result Format

2 Computational Units

2.3.2.2 Input Formats

To facilitate multiprecision multiplications, the multiplier accepts X and Y inputs represented in any combination of signed twos-complement format and unsigned format.

X input		Y input
signed	x	signed
unsigned	x	signed
signed	x	unsigned
unsigned	x	unsigned

The input formats are specified as part of the instruction. These are dynamically selectable each time the multiplier is used.

The (signed x signed) mode is used when multiplying two signed single precision numbers or the two upper portions of two signed multiprecision numbers.

The (unsigned x signed) and (signed x unsigned) modes are used when multiplying the upper portion of a signed multiprecision number with the lower portion of another or when multiplying a signed single precision number by an unsigned single precision number.

The (unsigned x unsigned) mode is used when multiplying unsigned single precision numbers or the non-upper portions of two signed multiprecision numbers.

2.3.2.3 MAC Input/Output Registers

The sources of MAC input and output are:

Source for X input port	Source for Y input port	Destination for R output port
MX0, MX1	MY0, MY1	MR (MR2, MR1, MR0)
AR	MF	MF
MR0, MR1, MR2		
SR0, SR1		

Computational Units   2

2.3.2.4   MR Register Operation

As described, and shown on the block diagram, the MR register is divided into three sections: MR0 (bits 0-15), MR1 (bits 16-31), and MR2 (bits 32-39). Each of these registers can be loaded from the DMD bus and output to the R bus or the DMD bus.

The 8-bit MR2 register is tied to the lower 8 bits of these buses. When MR2 is output onto the DMD bus or the R bus, it is sign extended to form a 16-bit value. MR1 also has an automatic sign-extend capability. When MR1 is loaded from the DMD bus, every bit in MR2 will be set to the sign bit (MSB) of MR1, so that MR2 appears as an extension of MR1. To load the MR2 register with a value other than MR1's sign extension, you must load MR2 after MR1 has been loaded. Loading MR0 affects neither MR1 nor MR2; no sign extension occurs in MR0 loads.

2.3.2.5   MAC Overflow And Saturation

The adder/subtracter generates an overflow status signal (MV) which is loaded into the processor arithmetic status (ASTAT) every time a MAC operation is executed. The MV bit is set when the accumulator result, interpreted as a twos-complement number, crosses the 32-bit (MR1/MR2) boundary. That is, MV is set if the upper nine bits of MR are not all ones or all zeros.

The MR register has a saturation capability which sets MR to the maximum positive or negative value if an overflow or underflow has occurred. The saturation operation depends on the overflow status bit (MV) in the processor arithmetic status (ASTAT) and the MSB of the MR2 register. The following table summarizes the MR saturation operation.

MV	MSB of MR2	MR content after saturation	
0	0 or 1	no change	
1	0	00000000 0111111111111111 1111111111111111	*full-scale positive*
1	1	11111111 1000000000000000 0000000000000000	*full-scale negative*

Table 2.3   Effect Of MAC Saturation Instruction

Saturation in the MAC is an instruction rather than a mode as in the ALU. The saturation instruction is intended to be used at the completion of a string of multiplication/accumulations so that intermediate overflows do not cause the accumulator to saturate.

2   Computational Units

Overflowing beyond the MSB of MR2 should never be allowed. The true sign bit of the result is then irretrievably lost and saturation may not produce a correct value. It takes more than 255 overflows (MV type) to reach this state, however.

2.3.2.6   Rounding Mode

The accumulator has the capability for rounding the 40-bit result R at the boundary between bit 15 and bit 16. Rounding can be specified as part of the instruction code. The rounded output is directed to either MR or MF. When rounding is invoked with MF as the output register, register contents in MF represent the rounded 16-bit result. Similarly, when MR is selected as the output, MR1 contains the rounded 16-bit result; the rounding effect in MR1 affects MR2 as well and MR2 and MR1 represent the rounded 24-bit result.

The accumulator uses an unbiased rounding scheme. The conventional method of biased rounding is to add a 1 into bit position 15 of the adder chain. This method causes a net positive bias since the midway value (when MR0=0x8000) is always rounded upward. The accumulator eliminates this bias by forcing bit 16 in the result output to zero when it detects this midway point. This has the effect of rounding odd MR1 values upward and even MR1 values downward, yielding a zero large-sample bias assuming uniformly distributed values.

Using x to represent any bit pattern (not all zeros), here are two examples of rounding. The first example is the typical rounding operation.

Example 1	*MR2*	*MR1*	*MR0*
Unrounded value:	xxxxxxxx	xxxxxxxx00100101	1xxxxxxxxxxxxxxx
Bit 15 = 1 Add 1 to bit 15 and carry			1
Rounded value:	xxxxxxxx	xxxxxxxx00100110	0xxxxxxxxxxxxxxx

The compensation to avoid net bias becomes visible when the lower 15 bits are all zero and bit 15 is one, i.e. the midpoint value.

Computational Units 2

Example 2	MR2	MR1	MR0

Example 2   MR2   MR1   MR0

Unrounded value: xxxxxxxx   xxxxxxxx01100110 1000000000000000

Bit 15 = 1 and bits 0-14 = 0
Add 1 to bit 15 and carry   1

                    xxxxxxxx   xxxxxxxx01100111 0000000000000000

Since bit 16 = 1, force it to 0

Rounded value: xxxxxxxx   xxxxxxxx01100110 0000000000000000

In this last case, bit 16 is forced to zero. This algorithm is employed on every rounding operation, but is only evident when the bit patterns shown in the lower 16 bits of the last example are present.

2.4   BARREL SHIFTER

The shifter provides a complete set of shifting functions for 16-bit inputs, yielding a 32-bit output. These include arithmetic shift, logical shift and normalization. The shifter also performs derivation of exponent and derivation of common exponent for an entire block of numbers. These basic functions can be combined to efficiently implement any degree of numerical format control, including full floating-point representation.

2.4.1   Shifter Block Diagram Discussion

Figure 2.9, on the next page, shows a block diagram of the shifter. The shifter can be divided into the following components: the shifter array, the OR/PASS logic, the exponent detector, and the exponent compare logic.

The shifter array is a 16x32 barrel shifter. It accepts a 16-bit input and can place it anywhere in the 32-bit output field, from off-scale right to off-scale left, in a single cycle. This gives 49 possible placements within the 32-bit field. The placement of the 16 input bits is determined by a control code (C) and a HI/LO reference signal.

The shifter array and its associated logic are surrounded by a set of registers. The shifter input (SI) register provides input to the shifter array and the exponent detector. The SI register is 16 bits wide and is readable and writable from the DMD bus. The shifter array and the exponent detector also take as inputs AR, SR or MR via the R bus. The shifter result

2 Computational Units

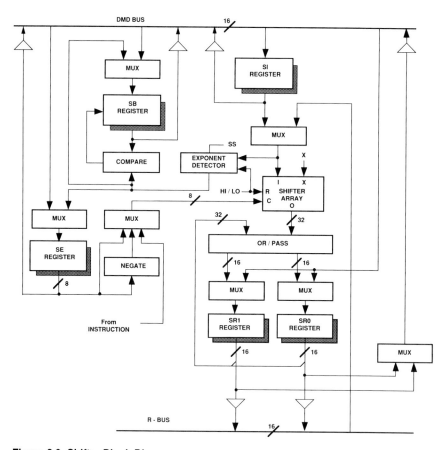

Figure 2.9 Shifter Block Diagram

(SR) register is 32 bits wide and is divided into two 16-bit sections, SR0 and SR1. The SR0 and SR1 registers can be loaded from the DMD bus and output to either the DMD bus or the R bus. The SR register is also fed back to the OR/PASS logic to allow double-precision shift operations.

The SE register ("shifter exponent") is 8 bits wide and holds the exponent during the normalize and denormalize operations. The SE register is loadable and readable from the lower 8 bits of the DMD bus. It is a two-complement, 8.0 value.

36

Computational Units   2

The SB register ("shifter block") is important in block floating-point operations where it holds the block exponent value, that is, the value by which the block values must be shifted to normalize the largest value. SB is 5 bits wide and holds the most recent block exponent value. The SB register is loadable and readable from the lower 5 bits of the DMD bus. It is a twos-complement, 5.0 value.

Whenever the SE or SB registers are output onto the DMD bus, they are sign-extended to form a 16-bit value.

Any of the SI, SE or SR registers can be read and written in the same cycle. Registers are read at the beginning of the cycle and written at the end of the cycle. All register reads, therefore, read values loaded at the end of a previous cycle. A new value written to a register cannot be read out until a subsequent cycle. This allows an input register to provide an operand to the shifter at the beginning of the cycle and be updated with the next operand at the end of the same cycle. It also allows a result register to be stored in memory and updated with a new result in the same cycle. See the discussion of "Multifunction Instructions" in Chapter 12, "Instruction Set Reference" for an illustration of this same-cycle read and write.

The shifter contains a duplicate bank of registers, shown in Figure 2.9 behind the primary registers. There are actually two sets of SE, SB, SI, SR1, and SR0 registers. Only one bank is accessible at a time. The additional bank of registers can be activated for extremely fast context switching. A new task, such as an interrupt service routine, can then be executed without transferring current states to storage.

The selection of the primary or alternate bank of registers is controlled by bit 0 in the processor mode status register (MSTAT). If this bit is a 0, the primary bank is selected; if it is a 1, the secondary bank is selected.

The shifting of the input is determined by a control code (C) and a HI/LO reference signal. The control code is an 8-bit signed value which indicates the direction and number of places the input is to be shifted. Positive codes indicate a left shift (upshift) and negative codes indicate a right shift (downshift). The control code can come from three sources: the content of the shifter exponent (SE) register, the negated content of the SE register or an immediate value from the instruction.

2 Computational Units

The HI/LO signal determines the reference point for the shifting. In the HI state, all shifts are referenced to SR1 (the upper half of the output field), and in the LO state, all shifts are referenced to SR0 (the lower half). The HI/LO reference feature is useful when shifting 32-bit values since it allows both halves of the number to be shifted with the same control code. HI/LO reference signal is selectable each time the shifter is used.

The shifter fills any bits to the right of the input value in the output field with zeros, and bits to the left are filled with the extension bit (X). The extension bit can be fed by three possible sources depending on the instruction being performed. The three sources are the MSB of the input, the AC bit from the arithmetic status register (ASTAT) or a zero.

Table 2.4 shows the shifter array output as a function of the control code and HI/LO signal.

The OR/PASS logic allows the shifted sections of a multiprecision number to be combined into a single quantity. When PASS is selected, the shifter array output is passed through and loaded into the shifter result (SR) register unmodified. When OR is selected, the shifter array is bitwise ORed with the current contents of the SR register before being loaded there.

The exponent detector derives an exponent for the shifter input value. The exponent detector operates in one of three ways which determine how the input value is interpreted. In the HI state, the input is interpreted as a single precision number or the upper half of a double precision number. The exponent detector determines the number of leading sign bits and produces a code which indicates how many places the input must be up-shifted to eliminate all but one of the sign bits. The code is negative so that it can become the effective exponent for the mantissa formed by removing the redundant sign bits.

In the HI-extend state (HIX), the input is interpreted as the result of an add or subtract performed in the ALU section which may have overflowed. Therefore the exponent detector takes the arithmetic overflow (AV) status into consideration. If AV is set, then a +1 exponent is output to indicate an extra bit is needed in the normalized mantissa (the ALU Carry bit); if AV is not set, then HI-extend functions exactly like the HI state. When performing a derive exponent function in HI or HI-extend modes, the exponent detector also outputs a shifter sign (SS) bit which is loaded into the arithmetic status register (ASTAT). The sign bit is the same as the

Computational Units 2

Control Code		Shifter Array Output			
HI reference	**LO Reference**				
+16 to +127	+32 to +127	00000000	00000000	00000000	00000000
+15	+31	R0000000	00000000	00000000	00000000
+14	+30	PR000000	00000000	00000000	00000000
+13	+29	NPR00000	00000000	00000000	00000000
+12	+28	MNPR0000	00000000	00000000	00000000
+11	+27	LMNPR000	00000000	00000000	00000000
+10	+26	KLMNPR00	00000000	00000000	00000000
+9	+25	JKLMNPR0	00000000	00000000	00000000
+8	+24	IJKLMNPR	00000000	00000000	00000000
+7	+23	HIJKLMNP	R0000000	00000000	00000000
+6	+22	GHIJKLMN	PR000000	00000000	00000000
+5	+21	FGHIJKLM	NPR00000	00000000	00000000
+4	+20	EFGHIJKL	MNPR0000	00000000	00000000
+3	+19	DEFGHIJK	LMNPR000	00000000	00000000
+2	+18	CDEFGHIJ	KLMNPR00	00000000	00000000
+1	+17	BCDEFGHI	JKLMNPR0	00000000	00000000
0	+16	ABCDEFGH	IJKLMNPR	00000000	00000000
-1	+15	XABCDEFG	HIJKLMNP	R0000000	00000000
-2	+14	XXABCDEF	GHIJKLMN	PR000000	00000000
-3	+13	XXXABCDE	FGHIJKLM	NPR00000	00000000
-4	+12	XXXXABCD	EFGHIJKL	MNPR0000	00000000
-5	+11	XXXXXABC	DEFGHIJK	LMNPR000	00000000
-6	+10	XXXXXXAB	CDEFGHIJ	KLMNPR00	00000000
-7	+9	XXXXXXXA	BCDEFGHI	JKLMNPR0	00000000
-8	+8	XXXXXXXX	ABCDEFGH	IJKLMNPR	00000000
-9	+7	XXXXXXXX	XABCDEFG	HIJKLMNP	R0000000
-10	+6	XXXXXXXX	XXABCDEF	GHIJKLMN	PR000000
-11	+5	XXXXXXXX	XXXABCDE	FGHIJKLM	NPR00000
-12	+4	XXXXXXXX	XXXXABCD	EFGHIJKL	MNPR0000
-13	+3	XXXXXXXX	XXXXXABC	DEFGHIJK	LMNPR000
-14	+2	XXXXXXXX	XXXXXXAB	CDEFGHIJ	KLMNPR00
-15	+1	XXXXXXXX	XXXXXXXA	BCDEFGHI	JKLMNPR0
-16	0	XXXXXXXX	XXXXXXXX	ABCDEFGH	IJKLMNPR
-17	-1	XXXXXXXX	XXXXXXXX	XABCDEFG	HIJKLMNP
-18	-2	XXXXXXXX	XXXXXXXX	XXABCDEF	GHIJKLMN
-19	-3	XXXXXXXX	XXXXXXXX	XXXABCDE	FGHIJKLM
-20	-4	XXXXXXXX	XXXXXXXX	XXXXABCD	EFGHIJKL
-21	-5	XXXXXXXX	XXXXXXXX	XXXXXABC	DEFGHIJK
-22	-6	XXXXXXXX	XXXXXXXX	XXXXXXAB	CDEFGHIJ
-23	-7	XXXXXXXX	XXXXXXXX	XXXXXXXA	BCDEFGHI
-24	-8	XXXXXXXX	XXXXXXXX	XXXXXXXX	ABCDEFGH
-25	-9	XXXXXXXX	XXXXXXXX	XXXXXXXX	XABCDEFG
-26	-10	XXXXXXXX	XXXXXXXX	XXXXXXXX	XXABCDEF
-27	-11	XXXXXXXX	XXXXXXXX	XXXXXXXX	XXXABCDE
-28	-12	XXXXXXXX	XXXXXXXX	XXXXXXXX	XXXXABCD
-29	-13	XXXXXXXX	XXXXXXXX	XXXXXXXX	XXXXXABC
-30	-14	XXXXXXXX	XXXXXXXX	XXXXXXXX	XXXXXXAB
-31	-15	XXXXXXXX	XXXXXXXX	XXXXXXXX	XXXXXXXA
-32 to -128	-16 to -128	XXXXXXXX	XXXXXXXX	XXXXXXXX	XXXXXXXX

ABCDEFGHIJKLMNPR represents the 16-bit input pattern

X stands for the extension bit

Table 2.4 Shifter Array Characteristic

2 Computational Units

MSB of the shifter input except when AV is set; when AV is set in HI-extend state, the MSB is inverted to restore the sign bit of the overflowed value.

In the LO state, the input is interpreted as the lower half of a double precision number. In the LO state, the exponent detector interprets the SS bit in the arithmetic status register (ASTAT) as the sign bit of the number. The SE register is loaded with the output of the exponent detector only if SE contains –15. This occurs only when the upper half–which must be processed first–contained all sign bits. The exponent detector output is also offset by –16 to account for the fact that the input is actually the lower half of a 32-bit value. Table 2.5 gives the exponent detector characteristics for all three modes.

The exponent compare logic is used to find the largest exponent value in an array of shifter input values. The exponent compare logic in conjunction with the exponent detector derives a block exponent. The comparator compares the exponent value derived by the exponent detector with the value stored in the shifter block exponent (SB) register and updates the SB register only when the derived exponent value is larger than the value in SB register. See the examples below.

2.4.2   Shifter Operations

The shifter performs the following functions (instruction mnemonics shown in parentheses):

- Arithmetic Shift   (ASHIFT)
- Logical Shift   (LSHIFT)
- Normalize   (NORM)
- Derive Exponent   (EXP)
- Block Exponent Adjust   (EXPADJ)

These basic shifter instructions can be used in a variety of ways, depending on the underlying arithmetic requirements. The following sections present single and multiple precision examples for these functions:

- Derivation of a Block Exponent
- Immediate Shifts
- Denormalization
- Normalization

Computational Units   2

S = Sign bit
N = Non-sign bit
D = Don't care bit

HI Mode

Shifter Array Input	Output
SNDDDDDD DDDDDDDD	0
SSNDDDDD DDDDDDDD	-1
SSSNDDDD DDDDDDDD	-2
SSSSNDDD DDDDDDDD	-3
SSSSSNDD DDDDDDDD	-4
SSSSSSND DDDDDDDD	-5
SSSSSSSN DDDDDDDD	-6
SSSSSSSS NDDDDDDD	-7
SSSSSSSS SNDDDDDD	-8
SSSSSSSS SSNDDDDD	-9
SSSSSSSS SSSNDDDD	-10
SSSSSSSS SSSSNDDD	-11
SSSSSSSS SSSSSNDD	-12
SSSSSSSS SSSSSSND	-13
SSSSSSSS SSSSSSSN	-14
SSSSSSSS SSSSSSSS	-15

HIX Mode

AV	Shifter Array Input	Output
1	DDDDDDDD DDDDDDDD	+1
0	SNDDDDDD DDDDDDDD	0
0	SSNDDDDD DDDDDDDD	-1
0	SSSNDDDD DDDDDDDD	-2
0	SSSSNDDD DDDDDDDD	-3
0	SSSSSNDD DDDDDDDD	-4
0	SSSSSSND DDDDDDDD	-5
0	SSSSSSSN DDDDDDDD	-6
0	SSSSSSSS NDDDDDDD	-7
0	SSSSSSSS SNDDDDDD	-8
0	SSSSSSSS SSNDDDDD	-9
0	SSSSSSSS SSSNDDDD	-10
0	SSSSSSSS SSSSNDDD	-11
0	SSSSSSSS SSSSSNDD	-12
0	SSSSSSSS SSSSSSND	-13
0	SSSSSSSS SSSSSSSN	-14
0	SSSSSSSS SSSSSSSS	-15

LO Mode

SS	Shifter Array Input	Output
S	NDDDDDDD DDDDDDDD	-15
S	SNDDDDDD DDDDDDDD	-16
S	SSNDDDDD DDDDDDDD	-17
S	SSSNDDDD DDDDDDDD	-18
S	SSSSNDDD DDDDDDDD	-19
S	SSSSSNDD DDDDDDDD	-20
S	SSSSSSND DDDDDDDD	-21
S	SSSSSSSN DDDDDDDD	-22
S	SSSSSSSS NDDDDDDD	-23
S	SSSSSSSS SNDDDDDD	-24
S	SSSSSSSS SSNDDDDD	-25
S	SSSSSSSS SSSNDDDD	-26
S	SSSSSSSS SSSSNDDD	-27
S	SSSSSSSS SSSSSNDD	-28
S	SSSSSSSS SSSSSSND	-29
S	SSSSSSSS SSSSSSSN	-30
S	SSSSSSSS SSSSSSSS	-31

Table 2.5 Exponent Detector Characteristics

2 Computational Units

The shift functions (arithmetic shift, logical shift, and normalize) can be optionally specified with PASS/OR and HI/LO modes so as to facilitate multiprecision operations. PASS passes the value through to SR directly. OR logically ORs the shift result with the current contents of SR. OR is used to join two 16-bit quantities into a 32-bit value in SR. The HI and LO modifiers reference the shift to the upper or lower half of the 32-bit SR register. These shift functions take inputs from either the SI register or any other result register and load the 32-bit shifted result into the SR register.

2.4.2.1 Shifter Input/Output Registers
The sources of shifter input and output are:

Source for Shifter input	Destination for Shifter output
SI	SR (SR0, SR1)
AR	
MR0, MR1, MR2	
SR0, SR1	

2.4.2.2 Derive Block Exponent
This function detects the exponent of the number largest in magnitude in an array of numbers. The EXPADJ instruction performs this function. The sequence of steps for a typical example is shown below.

A. Load SB with –16

The SB register is used to contain the exponent for the entire block. The possible values at the conclusion of a series of EXPADJ operations range from –15 to 0. The exponent compare logic updates the SB register if the new value is greater than the current value. Loading the register with –16 initializes it to a value certain to be less than any actual exponents detected.

B. Process the first array element:

Array(1) =   `11110101 10110001`

Exponent =   `-3`

– 3 > SB (–16)

SB gets   `-3`

42

Computational Units   2

C. *Process next array element:*

Array(2)=   `00000001 01110110`

Exponent =   `-6`

–6 < –3

SB remains   `-3`

D. *Continue processing array elements.*

When and if an array element is found whose exponent is greater than SB, that value is loaded into SB. When all array elements have been processed, the SB register contains the exponent of the largest number in the entire block. No normalization is performed. EXPADJ is purely an inspection operation. The value in SB could be transferred to SE and used to normalize the block on the next pass through the shifter. Or it could be simply associated with that data for subsequent interpretation.

2.4.2.3 *Immediate Shifts*
An immediate shift simply shifts the input bit pattern to the right (downshift) or left (upshift) by a given number of bits. Immediate shift instructions use the data value in the instruction itself to control the amount and direction of the shifting operation. (See the chapter "Instruction Set Overview" for an example of this instruction.) The data value controlling the shift is an 8-bit signed number. The SE register is not used or changed by an immediate shift.

The following example shows the input value downshifted relative to the upper half of SR (SR1). This is the (HI) version of the shift.

```
SI=0xB6A3;
SR=LSHIFT SI BY -5 (HI);
```

Input   `10110110 10100011`

Shift value   `-5`

SR   `00000101 10110101 00011 000 000000`

2 Computational Units

Here is the same input value shifted in the other direction, referenced to the lower half (LO) of SR.

```
SI=0xB6A3;
SR=LSHIFT SI BY 5 (LO);
```

Input   10110110 10100011

Shift value   +5

SR   00000000 000 **10110 11010100 011** 00000

In addition to the direction of the shifting operation, the shift may be either arithmetic (ASHIFT) or logical (LSHIFT). For example, the following shows a logical shift, relative to the upper half of SR (HI).

```
SI=0xB6A3;
SR=LSHIFT SI BY -5 (HI);
```

Input   10110110 10100011

Shift value   -5

SR   00000 **101 10110101 00011** 000 00000000

This example shows an arithmetic shift of the same input and shift code.

```
SI=0xB6A3;
SR=ASHIFT SI BY -5 (HI);
```

Input   10110110 10100011

Shift value   -5

SR   11111**101 10110101 00011** 000 00000000

Computational Units   2

2.4.2.4   *Denormalize*

Denormalizing refers to shifting a number according to a predefined exponent. The operation is effectively a floating-point to fixed-point conversion.

Denormalizing requires a sequence of operations. First, the SE register must contain the exponent value. This value may be explicitly loaded or may be the result of some previous operation. Next the shift itself is performed, taking its shift value from the SE register, not from an immediate data value.

There are two examples of denormalizing a double-precision number below. The first shows a denormalization in which the upper half of the number is shifted first, followed by the lower half. Since computations may produce output in either order, the second example shows the same operation in the other order, i.e. lower half first.

Always select the arithmetic shift for the higher half (HI) of the twos-complement input (or logical for unsigned). Likewise, the first half processed uses the PASS modifier.

Modifiers = HI, PASS   Shift operation = Arithmetic, SE = –3

First Input   10110110 10100011   (upper half of desired result)

SR   111**10110 11010100 011** 00000 00000000

Now the lower half is processed. Always select a logical shift for the lower half of the input. Likewise, the second half processed must use the OR modifier to avoid overwriting the previous half of the output value.

Modifiers = LO, OR   Shift operation = Logical, SE = –3

Second Input   01110110 01011101   (lower half of desired result)

SR   11110110 11010100 011 **01110 11001011**

2 Computational Units

Here is the same input processed in the reverse order. The higher half is always arithmetically shifted and the lower half is logically shifted. The first input is PASSed through to SR, but the second half is ORed to create one double-precision value in SR.

Modifiers = LO, PASS   Shift operation = Logical, SE = –3

First Input   01110110 01011101   (lower half of desired result)

SR   00000000 00000000 000 **01110 11001011**

Modifiers = HI, OR   Shift operation = Arithmetic, SE = –3

Second Input   10110110 10100011   (upper half of desired result)

SR   111**10110 11010100 011** 01110 11001011

2.4.2.5 Normalize

Numbers with redundant sign bits require normalizing. Normalizing a number is the process of shifting a twos-complement number within a field so that the rightmost sign bit lines up with the MSB position of the field and recording how many places the number was shifted. The operation can be thought of as a fixed-point to floating-point conversion, generating an exponent and a mantissa.

Normalizing is a two-stage process. The first stage derives the exponent. The second stage does the actual shifting. The first stage uses the EXP instruction which detects the exponent value and loads in into the SE register. This instruction (EXP) recognizes a (HI) and (LO) modifier. The second stage uses the NORM instruction. NORM recognizes (HI) and (LO) and the PASS and OR modifiers as well. NORM uses the negated value of the SE register as its shift control code. The negated value is used so that the shift is made in the correct direction.

Computational Units   2

Here is a normalization example for a single precision input.

```
SE=EXP AX0 (HI);
```

Detects Exponent With Modifier = HI

Input   `11110110 11010100`

SE set to   −3

Normalize, with modifier = HI Shift driven by value in SE

Input   `11110110 11010100`

SR   **`10110110 10100`**`000 00000000 00000000`

For a single precision input, the normalize operation can use either the (HI) or (LO) modifier, depending on whether you want the result in SR1 or SR0, respectively.

Double precision values follow the same general scheme. The first stage detects the exponent and the second stage normalizes the two halves of the input. For double precision, however, there are two operations in each stage.

For the first stage, the upper half of the input must be operated on first. This first exponent derivation loads the exponent value into SE. The second exponent derivation, operating on the lower half of the number will not alter the SE register unless SE = −15. This happens only when the first half contained all sign bits. In this case, the second operation will load a value into SE. (See Table 2.5) This value is used to control both parts of the normalization that follows.

For the second stage, now that SE contains the correct exponent value, the order of operations is immaterial. The first half (whether HI or LO) is normalized with the PASS modifier and the second half with the OR modifier to create one double precision value in SR. The (HI) and (LO) modifiers identify which half is being processed.

2   Computational Units

Here is a complete example of a typical double precision normalization.

1. *Detect Exponent, Modifier = HI*

   First Input   11110110 11010100   (Must be upper half)

   SE set to   -3

2. *Detect Exponent, Modifier = LO*

   Second Input   01101110 11001011

   SE unchanged, still -3

3. *Normalize, Modifiers = HI, PASS, SE = –3*

   First Input   11110110 11010100

   SR   **10110110 10100** 000 00000000 00000000

4. *Normalize , Modifiers = LO, OR, SE = –3*

   Second Input   01101110 11001011

   SR   10110110 10100 **011 01110110 01011** 000

If the upper half of the input contains all sign bits, the SE register value is determined by the second derive exponent operation as shown below.

1. *Detect Exponent, Modifier = HI*

   First Input   11111111 11111111   (Must be upper half)

   SE set to   -15

2. *Detect Exponent, Modifier = LO*

   Second Input   11110110 11010100

   SE now set to   -19

Computational Units   2

3. *Normalize, Modifiers = HI, PASS, SE = –19 (negated)*

First Input   11111111 11111111

SR   00000000 00000000 00000000 00000000

All values of SE less than –15 (resulting in a shift of +16 or more) upshift the input completely off scale.

4. *Normalize, Modifiers = LO, OR, SE = –19 (negated)*

Second Input   11110110 11010100

SR   **10110110 10100** 000 00000000 00000000

There is one additional normalization situation, requiring the HI-extended (HIX) state. This is specifically when normalizing ALU results (AR) that may have overflowed. This operation reads the arithmetic status word (ASTAT) overflow bit (AV) and the carry bit (AC) in conjunction with the value in AR. AV is set (1) if an overflow has occurred. AC contains the true sign of the twos-complement value.

For example, given these conditions:

AR =   11111010 00110010

AV =   1, indicating overflow

AC =   0, the true sign bit of this value

1. *Detect Exponent, Modifier = HIX*

SE gets set to   +1

2. *Normalize, Modifier = HI, SE = 1*

AR = 11111010 00110010

SR = **0**1111101 00011001

2 Computational Units

The AC bit is supplied as the sign bit, shown in bold above.

The HIX operation executes properly whether or not there has actually been an overflow. Consider this example.

AR =   11100011 01011011

AV =   0, indicating no overflow

AC =   0, not meaningful if AV = 0

1. *Detect Exponent, Modifier = HIX*

    SE set to   -2

2. *Normalize, Modifier = HI, SE = –2*

    AR = 11100011 01011011

    SR = **10001101 01101** 000 00000000 00000000

The AC bit is not used as the sign bit. A brief examination of Table 2.4 shows that the HIX mode is identical to the HI mode when AV is not set. When the NORM, LO operation is done, the extension bit is zero; when the NORM, HI operation is done, the extension bit is AC.

Program Control ■ 3

3.1 OVERVIEW

This chapter describes the program sequencer of the ADSP-2100 family processors. The program sequencer circuitry controls the flow of program execution. It contains an interrupt controller and status and condition logic. The instruction cache of the ADSP-2100 processor is described separately.

3.2 PROGRAM SEQUENCER

The program sequencer generates a stream of instruction addresses and provides flexible control of program flow. It allows sequential instruction execution, zero-overhead looping, sophisticated interrupt servicing, and single-cycle branching with jumps and calls (both conditional and unconditional).

Figure 3.1, which can be found on the following page, shows a block diagram of the program sequencer. Each functional block of the sequencer is discussed is detail in this chapter.

This chapter discusses both program sequencer logic and the following ADSP-2100 family instruction instructions used to control program flow:

DO UNTIL
JUMP
CALL
RTS *(Return From Subroutine)*
RTI *(Return From Interrupt)*
IDLE *(not implemented on ADSP-2100 processor)*
TRAP *(implemented only on ADSP-2100 processor)*

For a complete description of each instruction, refer to Chapter 12, "Instruction Set Reference."

3 Program Control

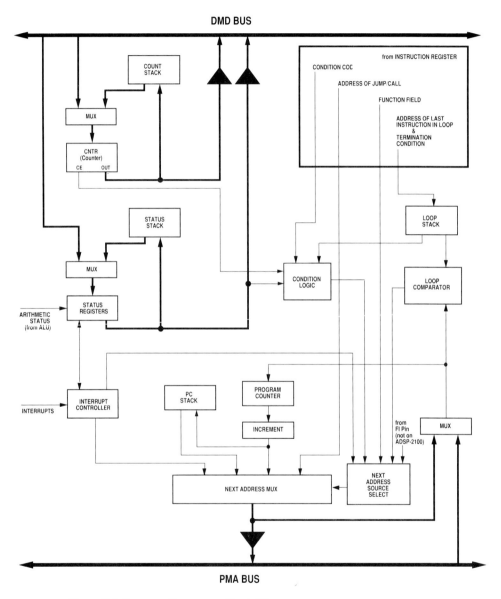

Figure 3.1 Program Sequencer Block Diagram

Program Control   3

3.2.1   Next Address Select Logic

While the processor is executing an instruction, the program sequencer pre-fetches the next instruction. The sequencer's next address select logic generates a program memory address (for the pre-fetch) from one of four sources:

• PC incrementer
• PC stack
• instruction register
• interrupt controller

The next address circuit (shown in Figure 3.1) selects which of these sources is used, based on inputs from the instruction register, condition logic, loop comparator and interrupt controller. The next instruction address is then output on the PMA bus for the pre-fetch.

The PC incrementer is selected as the source of the next address if program flow is sequential. This is also the case when a conditional jump or return is not taken and when a DO UNTIL loop terminates. The output of the PC incrementer is driven onto the PMA bus and is loaded back into the program counter to begin the next cycle.

The PC stack is used as the source for the next address when a return from subroutine or return from interrupt is executed. The top stack value is also used as the next address when returning to the top of a DO UNTIL loop.

The instruction register provides the next address when a direct jump is taken. The 14-bit jump address is embedded in the instruction word.

The interrupt controller provides the next program memory address when servicing an interrupt. Upon recognizing a valid interrupt, the processor jumps to the interrupt vector location corresponding to the active interrupt request.

Another possible source for the next address is one of the I4-I7 index registers of DAG2 (Data Address Generator 2), used when a register indirect jump is executed as in the following instruction:

```
JUMP (I4);
```

In this case the program counter (PC) is loaded from DAG2 via the PMA bus. (Data address generators are described in Chapter 4.)

3 Program Control

3.2.2 Program Counter And PC Stack

The program counter (PC) is a 14-bit register which always contains the address of the currently executing instruction. The output of the PC is fed into a 14-bit incrementer which adds 1 to the current PC value. The output of the incrementer can be selected by the next address multiplexer to fetch the next sequential instruction.

Associated with the PC is a 14-bit by 16-word stack that is pushed with the output of the incrementer when a CALL instruction is executed. The PC stack is also pushed when a DO UNTIL is executed and when an interrupt is processed. For interrupts, however, the incrementer is disabled so that the current PC value (instead of PC+1) is pushed. This allows the current instruction, which is aborted, to be refetched upon returning from the interrupt service routine. The pushing and popping of the PC stack occurs automatically in all of these cases. The stack can also be manually popped.

The output of the next address multiplexer is fed back to the PC, which normally reloads it at the end of each processor cycle. In the case of a register indirect jump, however, DAG2 drives the PMA bus with the next instruction address and the PC is loaded directly from the PMA bus.

3.2.3 Loop Counter And Stack

The counter and count stack provide the program sequencer with a powerful looping mechanism. The counter is a 14-bit register with automatic post-decrement capability that controls the flow of program loops which execute a predetermined number of times. Count values are 14-bit unsigned-magnitude values.

Before entering the loop, the counter (CNTR register) is loaded with the desired loop count from the lower 14 bits of the DMD bus. The actual loop count N is loaded, as opposed to $N–1$. This is due to the operation of the counter expired (CE) status logic, which tests CE (and automatically post-decrements the counter) at the end of a DO UNTIL loop that uses CE as its termination condition. CE is tested at the beginning of each processor cycle and the counter is decremented at the end; therefore CE is asserted when the counter reaches 1 so that the loop executes N times.

The counter may also be tested and automatically decremented by a conditional jump instruction that tests CE. The counter is not decremented when CE is checked as part of a conditional return or conditional arithmetic instruction.

Program Control   3

The counter may be read directly over the DMD bus at any time without affecting its contents. When reading the counter, the upper two bits of the DMD bus are padded with zeroes.

The count stack is a 14-bit by 4-word stack which allows nesting of loops by storing temporarily dormant loop counts. When a new value is loaded into the counter from the DMD bus, the current counter value is automatically pushed onto the count stack. The count stack is automatically popped whenever the CE status is tested and is true, thereby resuming execution of the outer loop (if any). The count stack may also be popped manually if an early exit from a loop is taken.

There are two exceptions to the automatic pushing of the count stack. A counter load from the DMD bus does not cause a count stack push if there is no valid value in the counter, because a stack location would be wasted on the invalid counter value. There is no valid value in the counter after a system reset and also after the CE condition is tested when the count stack is empty. The count stack empty status bit in the SSTAT register indicates when the stack is empty.

The second exception is provided explicitly by the special purpose syntax OWRCNTR (overwrite counter). Writing a value to OWRCNTR overwrites the counter with the new value, and nothing is pushed onto the count stack. OWRCNTR cannot be read (i.e. used as a source register). The OWRCNTR mnemonic can be used with all ADSP-2100 family processors except the ADSP-2100.

3.2.4   Loop Comparator And Stack

The DO UNTIL instruction initiates a zero-overhead loop using the loop comparator and loop stack of the program sequencer.

On every processor cycle, the loop comparator compares the next address generated by the program sequencer to the address of the last instruction of the loop (which is embedded in the DO UNTIL instruction). The address of the first instruction in the loop is maintained on the top of the PC stack. When the last instruction in the loop is executed the processor conditionally jumps to the beginning of the loop, eliminating the branching overhead otherwise incurred in loop execution.

The loop stack stores the last instruction addresses and termination conditions of temporarily dormant loops. Up to four levels can be stored. The only extra cycle associated with the nesting of DO UNTIL loops is the execution of the DO UNTIL instruction itself, since the pushing and

3 Program Control

popping of all stacks associated with the looping hardware is automatic. When using the counter expired (CE) status as the termination condition for the loop, another cycle is required for the initial loading of the counter. Table 3.1 shows the termination conditions that can be used with DO UNTIL.

Syntax	Status Condition	True If:
EQ	Equal Zero	AZ = 1
NE	Not Equal Zero	AZ = 0
LT	Less Than Zero	AN .XOR. AV = 1
GE	Greater Than or Equal Zero	AN .XOR. AV = 0
LE	Less Than or Equal Zero	(AN .XOR. AV) .OR. AZ = 1
GT	Greater Than Zero	(AN .XOR. AV) .OR. AZ = 0
AC	ALU Carry	AC = 1
NOT AC	Not ALU Carry	AC = 0
AV	ALU Overflow	AV = 1
NOT AV	Not ALU Overflow	AV = 0
MV	MAC Overflow	MV = 1
NOT MV	Not MAC Overflow	MV = 0
NEG	X Input Sign Negative	AS = 1
POS	X Input Sign Positive	AS = 0
CE	Counter Expired	
FOREVER	Always	

Table 3.1 DO UNTIL Termination Condition Logic

When a DO UNTIL instruction is executed, the 14-bit address of the last instruction and a 4-bit termination condition (both contained in the DO UNTIL instruction) are pushed onto the 18-bit by 4-word loop stack. Simultaneously, the PC incrementer output is pushed onto the PC stack. Since the DO UNTIL instruction is located just before the first instruction of the loop, the PC stack then contains the first loop instruction address, and the loop stack contains the last loop instruction address and termination condition. The non-empty state of the loop stack activates the loop comparator which compares the address on top of the loop stack with the address of the next instruction. When these two addresses are equal, the loop comparator notifies the next address source selector that the last instruction in the loop will be executed on the next cycle.

At this point, there are three possible results depending on the type of instruction at the end of the loop. Case 1 illustrates the most typical situation. Cases 2 and 3 are also allowed but involve greater program complexity for proper execution.

Program Control   3

Case 1
If the last instruction in the loop is not a jump, call, return, or idle, the next address circuit will select the next address based on the termination condition stored on the top of the loop stack. If the condition is false, the top address on the PC stack is selected, causing a fetch of the first instruction of the loop. If the termination condition is true, the PC incrementer is chosen, causing execution to fall out of the loop. The loop stack, PC stack, and counter stack (if being used) are then popped.

(Note that conditional arithmetic instructions execute based on the condition explicitly stated in the instruction, whereas the loop sequencing is controlled by the (implicit) termination condition contained on top of the stack.)

Case 2
If the last instruction in the loop is a jump, call, or return, the explicitly stated instruction takes precedence over the implicit sequencing of the loop. If the condition in the instruction is false, normal loop sequencing takes place as described for Case 1.

If the condition in the instruction is true, however, program control transfers to the jump/call/return address. Any actions that would normally occur upon an end-of-loop detection do not take place: fetching the first instruction of the loop, falling out of the loop and popping the loop stack, PC stack, and counter stack, or decrementing the counter.

(Note that for a return instruction, control is passed back to the top of the loop since the PC stack contains the beginning address of the loop.)

Case 3
If the last instruction in the loop is an IDLE, program flow is controlled by the IDLE instruction rather than the loop. When the IDLE instruction is executed, the processor enters a low-power wait-for-interrupt state. When the processor is interrupted, loop execution terminates and program execution continues with the first instruction following the loop.

Note: Caution is required when ending a loop with a JUMP, CALL, RETURN, or IDLE instruction, or when making a premature exit from a loop. Since none of the loop sequencing mechanisms are active while the

3 Program Control

jump/call/return is being performed, the loop, PC, and counter stacks are left with the looping information (since they are not popped). In this situation, a manual pop of each of the relevant stacks is required to restore the correct state of the processor. A subroutine call poses this problem only when it is the last instruction in a loop; in such cases, the return causes program flow to transfer to the instruction just after the loop. Calls within a loop that are not the last instruction operate as in Case 1.

The only restriction concerning DO UNTIL loops is that nested loops cannot terminate on the same instruction. Since the loop comparator can only check for one loop termination at a time, falling out of an inner loop by incrementing the PC would go beyond the end address of the outer loop if they terminated on the same instruction.

3.3   PROGRAM CONTROL INSTRUCTIONS

The following sections describe the primary instructions used to control program flow.

3.3.1   JUMP Instruction

The 14-bit jump address is embedded in the JUMP instruction word. When a JUMP instruction is decoded, the jump address is input directly to the next address mux of the program sequencer. The address is driven onto the PMA bus and fed back to the PC for the next cycle. The following instruction, for example,

```
JUMP fir_start;
```

jumps to the address of the label `fir_start`.

3.3.1.1   Register Indirect JUMPs

In this case of register indirect jumps, the jump address is supplied by one of the I registers of DAG2 (I4, I5, I6, or I7). (Data address generators are described in Chapter 4.) The address is driven onto the PMA bus by DAG2, and is loaded into the PC on the next cycle. For example, the instruction

```
JUMP (I4);
```

will jump to the address contained in the I4 register.

Program Control   3

3.3.2   CALL Instruction

The CALL instruction executes in a similar fashion as the JUMP instruction. The address of the subroutine is embedded in the CALL instruction word and, once extracted from the instruction register, is fed back the PC for the next cycle. In addition, the current value of the program counter is incremented and pushed onto the PC stack. Upon return from the subroutine, the PC stack is popped into the program counter and execution resumes with the instruction following the CALL.

3.3.3   DO UNTIL Loops

The most common form of a DO UNTIL loop uses the counter register (CNTR) as a loop iteration counter. When the counter is used to control loop iteration, CE (counter expired) must be used as the DO UNTIL termination condition. A simple example of this type of loop is as follows:

```
L0=10;               {setup circular buffer length register}
I0=^data_buffer;     {load pointer with first address of}
                     {circular buffer}
M0=1;                {setup modify register for pointer increment}
CNTR=10;             {load counter with circular buffer length}

DO loop UNTIL CE;    {repeat loop until counter expired}
   DM(I0,M0)=0;      {initialize/clear circular buffer}
   ...any instruction...
loop: ...any instruction...
```

When the

```
CNTR=10;
```

instruction is executed, prior to entering the loop, the counter is loaded via the DMD bus. Any previously existing count would be simultaneously pushed onto the count stack; this push operation is omitted if the counter is empty. The

```
DO loop UNTIL CE;
```

instruction itself only sets up the conditions for looping; no other operation occurs while the instruction is executed. This occurs only once, at the beginning of the first time through the loop.

3   Program Control

Execution of the DO UNTIL instruction pushes the address of the
instruction immediately following the DO UNTIL onto the PC stack (by
pushing the incremented PC). On the same cycle, the loop stack is pushed
with the address of the end-of-loop instruction and the termination
condition.

As execution continues within the loop, the loop comparator checks each
instruction's address against the address of the loop's last instruction.
Until that address is reached, normal execution continues.

Each time the end of the loop is reached, the loop comparator determines
that the currently executing instruction is the last in the loop. This affects
the next address select logic of the program sequencer: instead of using
the incremented PC for the next address, the loop termination condition is
evaluated. If the termination condition is false, execution continues with
the first instruction of the loop (the top of the PC stack is taken as the next
address). Note that the PC and loop stacks are not popped, only read.

On the final pass through the loop, the termination condition is true. The
PC stack is popped and execution continues with the instruction
immediately following the last instruction of the loop. The loop stack and
count stack are also popped on this cycle.

3.3.4   IDLE Instruction

The IDLE instruction causes the processor to wait indefinitely in a low
power state until an interrupt occurs. When an unmasked interrupt
occurs, it is serviced; execution then continues with the instruction
following the IDLE instruction. (The IDLE instruction is not implemented
on the ADSP-2100 processor.)

3.4   INTERRUPT CONTROLLER

The program sequencer's interrupt controller responds to interrupts by
shifting control to the instruction located at the appropriate interrupt
vector address. Tables 3.2–3.6 show the interrupts and associated vector
addresses for each processor of the ADSP-2100 family. (Note that SPORT1
can be configured as either a serial port or as a collection of control pins
including two external interrupt inputs, $\overline{IRQ0}$ and $\overline{IRQ1}$. See Chapter 5,
"Serial Ports," for more information about the configuration of SPORT1.)

Program Control   3

In the ADSP-2101, ADSP-2105, ADSP-2111, and ADSP-21msp50
processors, the interrupt vector locations are spaced four program
memory locations apart—this allows short interrupt service routines to be
coded in place, with no jump to the service routine required.

In the ADSP-2100 (and for interrupt service routines with more than four
instructions on the ADSP-2101, ADSP-2105, ADSP-2111, and ADSP-
21msp50), program control must be transferred to the service routine by
means of a jump instruction placed at the interrupt vector location.

After an interrupt has been serviced, an RTI (Return From Interrupt)
instruction returns control to the main program by popping the top value
on the PC stack into the PC; the status stack is also popped to restore the
previous processor state.

Interrupts can also be forced under software control, in all processors
except the ADSP-2100; see the discussion of the IFC register below.

Because of the efficient stack and program sequencer, there is no latency
(beyond synchronization delay) when processing unmasked interrupts,
even when interrupting DO UNTIL loops. Nesting of interrupts allows
higher-priority interrupts to interrupt any lower-priority interrupt service
routines that may currently be executing, also with no additional latency.

The ADSP-2100 family processors include a secondary register set which
can be used to provide a fresh set of ALU, MAC, and Shifter registers
during interrupt servicing. This feature allows single-cycle context
switching. Use of the secondary registers is described in the "Mode Status
Register (MSTAT)" section of this chapter.

Interrupt Source	Interrupt Vector Address
$\overline{IRQ0}$	0x0000
$\overline{IRQ1}$	0x0001
$\overline{IRQ2}$	0x0002
$\overline{IRQ3}$	0x0003
Program startup at $\overline{RESET}$	0x0004

Table 3.2 ADSP-2100 Interrupts & Interrupt Vector Addresses

3 Program Control

Interrupt Source	Interrupt Vector Address
Program startup at RESET	0x0000
IRQ2	0x0004 *(highest priority)*
SPORT0 Transmit	0x0008
SPORT0 Receive	0x000C
SPORT1 Transmit / IRQ1	0x0010
SPORT1 Receive / IRQ0	0x0014
Timer	0x0018 *(lowest priority)*

Table 3.3 ADSP-2101 Interrupts & Interrupt Vector Addresses

Interrupt Source	Interrupt Vector Address
Program startup at RESET	0x0000
IRQ2	0x0004 *(highest priority)*
SPORT1 Transmit / IRQ1	0x0010
SPORT1 Receive / IRQ0	0x0014
Timer	0x0018 *(lowest priority)*

Table 3.4 ADSP-2105 Interrupts & Interrupt Vector Addresses

Interrupt Source	Interrupt Vector Address
Program startup at RESET	0x0000
IRQ2	0x0004 *(highest priority)*
HIP Write from Host	0x0008
HIP Read to Host	0x000C
SPORT0 Transmit	0x0010
SPORT0 Receive	0x0014
SPORT1 Transmit / IRQ1	0x0018
SPORT1 Receive / IRQ0	0x001C
Timer	0x0020 *(lowest priority)*

Table 3.5 ADSP-2111 Interrupts & Interrupt Vector Addresses

Program Control   3

Interrupt Source	Interrupt Vector Address
Program startup at RESET	0x0000
Powerdown	0x002C
IRQ2	0x002C *(highest priority)*
HIP Write from Host	0x0008
HIP Read to Host	0x000C
SPORT0 Transmit	0x0010
SPORT0 Receive	0x0014
Analog (DAC) Transmit	0x0018
Analog (ADC) Receive	0x001C
SPORT1 Transmit / IRQ1	0x0020
SPORT1 Receive / IRQ0	0x0024
Timer	0x0028 *(lowest priority)*

Table 3.6 ADSP-21msp50 Interrupts & Interrupt Vector Addresses

3.4.1   Interrupt Servicing Sequence

When an interrupt request occurs, it is latched while the processor finishes executing the current instruction. The interrupt request is then compared with the interrupt mask register, IMASK, by the interrupt controller.

If the interrupt is not masked, the program sequencer pushes the current value of the program counter (which contains the address of the next instruction) onto the PC stack—this allows execution to continue, after the interrupt is serviced, with the next instruction of the main program. The program sequencer also pushes the current values of the ASTAT, MSTAT, and IMASK registers onto the status stack. ASTAT, MSTAT and IMASK are stored in this order, with the MSB of ASTAT first, and so on. When IMASK is pushed, it is automatically reloaded with a new value that determines whether or not interrupt nesting is allowed (based on the value of the interrupt nesting enable bit in ICNTL).

The processor then executes a NOP while simultaneously fetching the instruction located at the interrupt vector address. Upon return from the interrupt service routine, the PC and status stacks are popped and execution resumes with the next instruction of the main program.

3   Program Control

3.4.2   Configuring Interrupts

The following registers are used to configure interrupts:

- ICNTL—Configures each external ($\overline{IRQ}$x) interrupt as either edge-sensitive or level-sensitive; determines whether interrupts can be nested.
- IMASK—Masks or enables each individual interrupt (external and internal).
- IFC—Forces an interrupt or clears a pending edge-sensitive interrupt. *(Not implemented on ADSP-2100)*

For edge-sensitive $\overline{IRQ}$x interrupts on the ADSP-2101, ADSP-2105, ADSP-2111, and ADSP-21msp50, an interrupt request is latched internally whenever a falling edge (high-to-low transition) occurs at the interrupt input pin. The latch remains set until the interrupt is serviced; it is then automatically cleared. A pending edge-sensitive interrupt can also be cleared in software by setting the corresponding clear bit in the IFC register.

For edge-sensitive $\overline{IRQ}$x interrupts on the ADSP-2100, the processor detects an interrupt request by comparing the state of the interrupt input pin on successive clock cycles. When the ADSP-2100 detects an interrupt request signal which is inactive on one cycle and active the next, the interrupt request is latched internally. The latch remains set until the interrupt is serviced; it is then automatically cleared. For the ADSP-2100, therefore, interrupt request signals must be held active for at least one full cycle.

Edge-sensitive interrupt inputs generally require less external hardware than level-sensitive inputs, and allow signals such as sampling-rate clocks to be used as interrupts.

A level-sensitive interrupt must remain asserted until the interrupt is serviced. The interrupting device must then deassert the interrupt request so that the interrupt is not serviced again. Level-sensitive inputs, however, allow many interrupt sources to use the same input by combining them logically to produce a single interrupt request. Level-sensitive interrupts are not latched.

Software can also determine whether or not interrupts can be nested. In non-nesting mode, all interrupt requests are automatically masked out when an interrupt service routine is entered. In nesting mode, the processor allows higher-priority interrupts to be recognized and serviced.

64

Program Control   3

There are two levels of masking for the Host Interface Port (HIP) interrupts of the ADSP-2111 and ADSP-21msp50. The memory-mapped HMASK register configures masking out the generation of individual read or write interrupts for each HIP data register. The IMASK register can be set to mask or enable the servicing of all HIP read interrupts or all HIP write interrupts. Both IMASK and HMASK must be set for HDR interrupts. See Chapter 7, "Host Interface Port," for details.

3.4.2.1 Interrupt Control Register (ICNTL)

ICNTL is a 5-bit register that configures the external interrupt requests ($\overline{IRQx}$) of each processor. All bits in ICNTL are undefined after a processor reset. The bit definitions for each processor's ICNTL register are given in Appendix E, "Control/Status Registers."

ICNTL contains an $\overline{IRQx}$ sensitivity bit for each external interrupt. The sensitivity bits determine whether a given interrupt input is edge- or level-sensitive (0 = level-sensitive, 1 = edge-sensitive). There are no sensitivity bits for internally generated interrupts.

The interrupt nesting enable bit (bit 4) in ICNTL determines whether nesting of interrupt service routines is allowed.

3.4.2.2 Interrupt Mask Register (IMASK)

Each bit of the IMASK register enables or disables the servicing of an individual interrupt. If an edge-sensitive interrupt request signal occurs when the interrupt is masked, the request is latched but not serviced; the interrupt can then be recognized in software and serviced later. Specific bit definitions for each processor's IMASK register are given in Appendix E, "Control/Status Registers."

The contents of IMASK are automatically pushed onto the status stack when entering an interrupt service routine and popped back when returning from the routine. The configuration of IMASK upon entering the interrupt service routine is determined by the interrupt nesting enable bit (bit 4) of ICNTL; it may be altered, though, as part of the interrupt service routine itself.

When nesting is disabled, all interrupt levels are masked automatically (IMASK set to zero) when an interrupt service routine is entered.

3 Program Control

When nesting is enabled on the ADSP-2101, ADSP-2105, ADSP-2111, and ADSP-21msp50 processors, IMASK is set so that only equal and lower priority interrupts are masked; higher priority interrupts remain configured as they were prior to the interrupt. This is shown graphically, for the ADSP-2101, in Table 3.7.

When nesting is enabled on the ADSP-2100, IMASK is set so that equal and lower priority interrupts are masked. Higher priority interrupts, however, are automatically enabled. This is shown graphically in Table 3.8.

The mask bits are positive sense (0=masked, 1=enabled). IMASK is set to zero upon a processor reset. The interrupt nesting enable bit (in ICNTL) determines the state of IMASK upon entering the interrupt, as shown in Table 3.7 for the ADSP-2101.

ICNTL Interrupt Nesting Enable bit = 0 (nesting disabled)

Interrupt level serviced	IMASK contents before (pushed on stack)	IMASK contents entering interrupt service routine
0 (low)	ijklmn	000000
1	ijklmn	000000
2	ijklmn	000000
3	ijklmn	000000
4	ijklmn	000000
5 (high)	ijklmn	000000

ICNTL Interrupt Nesting Enable bit = 1 (nesting enabled)

Interrupt level serviced	IMASK contents before (pushed on stack)	IMASK contents entering interrupt service routine
0 (low)	ijklmn	ijklm0
1	ijklmn	ijkl00
2	ijklmn	ijk000
3	ijklmn	ij0000
4	ijklmn	i00000
5 (high)	ijklmn	000000

("ijklmn" represents any pattern of ones and zeroes)

Table 3.7 IMASK Entering Interrupt Service Routines (ADSP-2101)

ICNTL Interrupt Nesting Enable bit = 0 (nesting disabled)

Interrupt level serviced	IMASK contents before (pushed on stack)	IMASK contents entering interrupt service routine
0 (low)	ijkl	0000
1	ijkl	0000
2	ijkl	0000
3 (high)	ijkl	0000

ICNTL Interrupt Nesting Enable bit = 1 (nesting enabled)

Interrupt level serviced	IMASK contents before (pushed on stack)	IMASK contents entering interrupt service routine
0 (low)	ijkl	1110
1	ijkl	1100
2	ijkl	1000
3 (high)	ijkl	0000

("ijklmn" represents any pattern of ones and zeroes)

Table 3.8 IMASK Entering Interrupt Service Routines (ADSP-2100)

3.4.2.3 Interrupt Force & Clear Register (IFC)

IFC is a write-only register that allows the forcing and clearing of edge-sensitive interrupts in software. An interrupt is forced under program control by setting the force bit corresponding to the desired interrupt.

Edge-sensitive interrupts can be forced by setting the appropriate force bit in IFC. This causes the interrupt to be serviced once, unless masked. An external interrupt must be edge-sensitive (as determined by ICNTL) to be forced. The timer, SPORT, and analog ADC/DAC interrupts also behave like edge-sensitive interrupts and can be masked, cleared and forced.

Pending edge-sensitive interrupts can be cleared by setting the appropriate clear bit in IFC. Edge-triggered interrupts are cleared automatically when the corresponding interrupt service routine is called.

3 Program Control

Specific bit definitions for each processor's IFC register are given in Appendix E, "Control/Status Registers." The IFC register is not implemented in the ADSP-2100 processor. The IFC registers of the ADSP-2111 and ADSP-21msp50 processors do not include force/clear bits for Host Interface Port interrupts; HIP interrupts cannot be forced or cleared in software.

3.4.3   Interrupt Latency

For the $\overline{IRQ2}$, $\overline{IRQ1}$, $\overline{IRQ0}$, SPORT, HIP, and ADC/DAC interrupts of the ADSP-2101, ADSP-2105, ADSP-2111, and ADSP-21msp50 processors, the latency from when an interrupt occurs to when the first instruction of the service routine is executed is at least three full cycles. This is shown in Figure 3.2. Two cycles are required to synchronize the interrupt internally, assuming that setup and hold times are met (for $\overline{IRQ2}$, $\overline{IRQ1}$, $\overline{IRQ0}$).

Since interrupts are only serviced on instruction boundaries, the instruction(s) executed during these two cycles must be fully completed, including any extra cycles inserted due to Bus Request/Bus Grant or memory wait states, before execution continues.

The third cycle of latency is needed to fetch the first instruction stored at the interrupt vector location. During this cycle, the processor executes a NOP instead of the instruction that would normally have been executed. On the next cycle, execution continues at the first instruction of the interrupt service routine. The address of the aborted instruction is pushed onto the PC stack; it will be fetched when the interrupt service routine is completed.

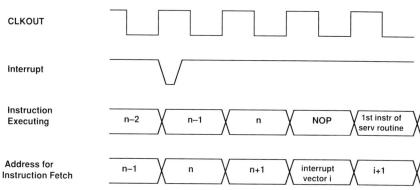

Figure 3.2 Interrupt Latency ($\overline{IRQ2-0}$, SPORT, HIP, & ADC/DAC Interrupts)

68

Program Control 3

For the timer interrupt of the ADSP-2101, ADSP-2105, ADSP-2111, and ADSP-21msp50 processors, the latency from when the interrupt occurs to when the first instruction of the service routine is executed is only one cycle. This is shown in Figure 3.3. The single cycle of latency is needed to fetch the instruction stored at the interrupt vector location.

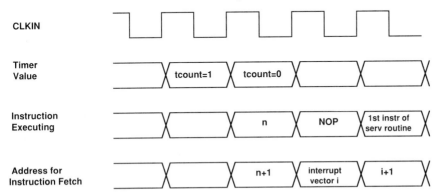

Figure 3.3 Interrupt Latency (Timer Interrupt)

For the IRQ3, IRQ2, IRQ1, and IRQ0 interrupts of the ADSP-2100, the latency from when an interrupt occurs to when the first instruction of the service routine is executed is two full cycles. This is shown in Figure 3.4, which can be found on the following page. One cycle is required to synchronize the interrupt internally, assuming that setup and hold times are met. The instruction executed during this cycle must be fully completed, including any extra cycles inserted due to Bus Request/Bus Grant, DMACK, or TRAP/HALT, before execution continues. The second cycle of latency is needed to fetch the instruction stored at the interrupt vector location.

3 Program Control

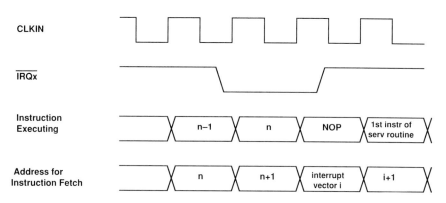

Figure 3.4 Interrupt Latency (ADSP-2100 Interrupts)

3.5 STATUS REGISTERS & STATUS STACK

Processor status and mode bits are maintained in internal registers which can be independently read and written over the DMD bus. These registers are:

ASTAT   Arithmetic status register
SSTAT   Stack status register *(read-only)*
MSTAT   Mode status register
ICNTL   Interrupt control register
IMASK   Interrupt mask register
IFC   Interrupt force/clear register *(write-only; not implemented in ADSP-2100)*

The interrupt-configuring status registers are described in the previous section. ASTAT, SSTAT, and MSTAT are discussed in the following sections.

The current ASTAT, MSTAT, and IMASK values are pushed onto the status stack when the processor responds to an interrupt; they are popped upon return from the interrupt service routine (with the RTI instruction). The depth of the stack varies from processor to processor. In each case, sufficient stack depth is provided to accommodate nesting of all interrupts.

Program Control   3

3.5.1   Arithmetic Status Register (ASTAT)

ASTAT is eight bits wide and holds the status information generated by the computational blocks of the processor. The individual bits of ASTAT are defined as shown in Figure 3.5. The bits which express a particular condition (AZ, AN, AV, AC, MV) are all positive sense (1=true, 0=false).

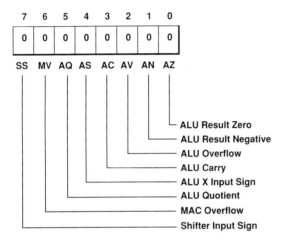

Figure 3.5 ASTAT Register

Each of the bits is automatically updated when a new status is generated by an arithmetic instruction. Each bit is affected only by a subset of arithmetic operations, as defined by the following table:

Status Bit	Updated by
AZ, AN, AV, AC	Any ALU operation except DIVS, DIVQ
AS	ALU absolute value operation (ABS)
AQ	ALU divide operations (DIVS, DIVQ)
MV	Any MAC operation except saturate MR (SAT MR)
SS	Shifter EXP operation

Arithmetic status is latched into ASTAT at the end of the cycle in which it was generated, and cannot be used until the next cycle.

Loading any ALU, MAC, or Shifter input or output registers directly from the DMD bus does not affect any of the arithmetic status bits. Executing the ALU instruction PASS sets the AZ and AN bits for a given X or Y operand and clears AC.

3   Program Control

3.5.2   Stack Status Register (SSTAT)

The SSTAT register is eight bits wide and holds information about the four processor stacks. The individual bits of SSTAT are defined as shown in Figure 3.6. All of the bits are positive sense (1=true, 0=false).

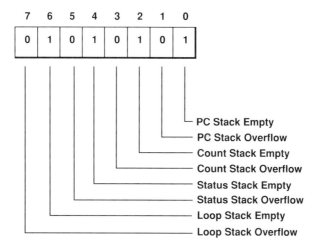

Figure 3.6 SSTAT Register (Read-Only)

The empty status bits indicate that the number of pop operations for the stack is greater than or equal to the number of push operations that have occurred since the last processor reset. The overflow status bits indicate that the number of push operations for the stack has exceeded the number of pop operations, by an amount that is greater than the total depth of the stack. When this occurs, the values most recently pushed will be missing from the stack—older stack values are considered more important than new.

Since a stack overflow represents a permanent loss of information, the stack overflow status bits "stick" once they are set, and subsequent pop operations have no effect on them. In this situation, then, it is possible to have both the stack empty and stack overflow bits set for a given stack.

Assume, for example, that the four-location count stack is overflowed by five successive pushes. Five successive pops will restore the stack empty condition, but will not clear the overflow condition. The processor must be reset to clear the stack overflow status.

Program Control   3

3.5.3   Mode Status Register (MSTAT)

The MSTAT register determines the operating mode of the processor. The individual bits of MSTAT are defined as shown in Figure 3.7.

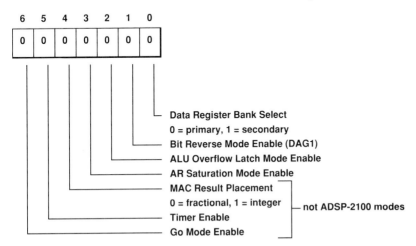

Figure 3.7 MSTAT Register

MSTAT can be modified by writing a new value to it with a MOVE instruction. Unlike the other status registers, MSTAT can also be altered with the Mode Control instruction (ENA, DIS). The Mode Control instruction provides a high-level, self-documenting method of configuring the processors' operating modes. Refer to the description of the Mode Control instruction in Chapter 12, "Instruction Set Reference," for further details.

To enable the bit reverse mode, for example, the following instruction could be used:

```
ENA BIT_REV;
```

The bit-reverse mode, when enabled, bitwise reverses all addresses generated by data address generator 1 (DAG1). This is useful for reordering the input or output data of an FFT algorithm.

3   Program Control

The ADSP-2100 family processors include a secondary register set which can be used to provide a fresh set of ALU, MAC, and Shifter registers at any time, for example during execution of a subroutine. The data register bank select bit of MSTAT determines which set of data registers is active (0=primary, 1=secondary). The secondary register set duplicates all of the input and result registers of the computation units, ALU, MAC, and Shifter:

AX0	MX0	SI
AX1	MX1	SE
AY0	MY0	SB
AY1	MY1	SR1
AF	MF	SR0
AR	MR0	
	MR1	
	MR2	

The following mode control instruction, for example, switches from the processor's primary register set to its secondary register set:

```
ENA SEC_REG;
```

while the following instruction switches back to the primary register set:

```
DIS SEC_REG;
```

The ALU overflow latch mode causes the AV status bit to "stick" once it is set. In this mode, AV will be set by an overflow and will remain set even if subsequent ALU operations do not generate overflows. AV can then be cleared only by writing a zero into it.

AR saturation mode, when enabled, causes AR to be saturated to the maximum positive (0x7FFF) or negative (0x8000) values whenever an ALU overflow occurs.

The MAC result placement mode determines whether the multiplier operates in integer or fractional format. This mode is discussed in Chapter 2, "Computational Units." (MAC result placement mode is not implemented on the ADSP-2100.)

Setting the timer enable bit causes the timer to begin decrementing. Clearing this bit halts the timer. (The timer enable bit is not implemented in the ADSP-2100.)

Program Control   3

Enabling GO mode allows the processor to continue executing instructions from internal program memory during a bus grant. The processor will halt, waiting for the buses to be released, only when an access of external memory is required. When GO mode is disabled, the processor always halts during bus grant. (GO mode is not implemented on the ADSP-2100.)

3.6   CONDITIONAL INSTRUCTIONS

The condition logic circuit of the program sequencer determines whether a conditional instruction is executed, for example a jump, call, or arithmetic operation. It also controls implicit loop sequencing operations based upon the loop continuation condition on top of the loop stack. The condition logic takes raw status information from ASTAT and the down counter and derives a set of sixteen composite status conditions.

The status conditions and corresponding assembly language syntax are listed in Table 3.9. These status conditions are used with the *IF condition* clause available on some instructions. In addition, the status of the FI pin (Flag In) can also be used as a condition for JUMP and CALL instructions (for all processors except the ADSP-2100).

Syntax	*Status Condition*	*True If:*
EQ	Equal Zero	AZ = 1
NE	Not Equal Zero	AZ = 0
LT	Less Than Zero	AN .XOR. AV = 1
GE	Greater Than or Equal Zero	AN .XOR. AV = 0
LE	Less Than or Equal Zero	(AN .XOR. AV) .OR. AZ = 1
GT	Greater Than Zero	(AN .XOR. AV) .OR. AZ = 0
AC	ALU Carry	AC = 1
NOT AC	Not ALU Carry	AC = 0
AV	ALU Overflow	AV = 1
NOT AV	Not ALU Overflow	AV = 0
MV	MAC Overflow	MV = 1
NOT MV	Not MAC Overflow	MV = 0
NEG	X Input Sign Negative	AS = 1
POS	X Input Sign Positive	AS = 0
NOT CE	Not Counter Expired	—
FLAG_IN*	FI pin	Last sample of FI pin = 1
NOT FLAG_IN*	Not FI pin	Last sample of FI pin = 0

* Only available on JUMP and CALL instructions.

Table 3.9 IF Condition Logic

75

3 Program Control

3.7   ADSP-2100 INSTRUCTION CACHE

The ADSP-2100 includes an on-chip instruction cache, a feature unique to this processor. The instruction cache stores sixteen previously executed instructions. When an instruction requires a data fetch from program memory, which would conflict with the instruction fetch, the cache, if valid, is used as the source of the instruction. Cache operation is transparent. No maintenance or overhead is required for either the storage or use of instructions in cache memory.

3.7.1   Cache Memory Operation

The cache is a 24-bit by 16-word memory array. While this stores a relatively small number of instructions, the multifunction instructions of the ADSP-2100 family instruction set allow a wide variety of algorithms to be coded within 16 locations. The cache is especially efficient for tightly looped algorithms such as digital filters and FFTs.

The cache interacts with the ADSP-2100's program sequencer and instruction register. Cache operation is as follows:

1. During normal operation the ADSP-2100 fetches the *(N+1)th* instruction while executing the *Nth* instruction. Each instruction fetched for the instruction register is also written into the cache. It is stored at the cache memory address specified by the four LSBs of the program memory address.

2. When the PMD bus is busy with a data transfer, the instruction register is loaded from cache. Note that, at this point, the validity of the loaded instruction has not been determined.

3. If the loaded instruction is valid, it is executed on the next cycle. If the instruction is not valid, the instruction register is cleared. An additional cycle is now required to fetch the next instruction. Validity is determined by the cache memory monitor described below.

When data can be read from program memory, the ADSP-2100 becomes, in effect, a processor with two data busses. For the multiply/accumulate operations typical of digital signal processing algorithms, this gives significant speed advantages. For program loops that can be stored completely in the cache, an additional cycle penalty is incurred only on the first pass through the loop. Once all instructions are resident in cache, the ADSP-2100 can simultaneously fetch two data values (using the DMD and PMD buses) and one instruction (from the cache).

Program Control   3

3.7.2   Cache Memory Monitor

The cache memory monitor logic keeps track of the program memory address range currently stored in the cache. One register contains the number of instructions that are valid ahead of the currently executing instruction, while another register contains the number of instructions that are valid behind the currently executing instruction. Cache addressing uses only the four LSBs of each program memory address.

Because the cache memory monitor can only follow the execution of instructions that are contiguous in memory, not all of the 16 cached instructions are necessarily valid. DO UNTIL loops and JUMPs contained entirely inside the cache allow efficient execution from the cache. A jump to an address outside the cache invalidates the entire cache. The number of valid instructions in the cache then increases until the cache fills or the program takes another out-of-cache jump. The cache size (as compared to loop size) is one limit to keep in mind when writing programs with program memory data transfers inside the loop.

Once the cache fills, newly fetched instructions are written over the oldest instructions in a circular manner (with modulo-16 addressing).

3.7.3   Programmers' Guidelines For Cache Memory

To allow programs to take full advantage of the ADSP-2100's on-chip instruction cache, the following constraints should be followed:

- The cache can contain no more than sixteen instructions: loops should fit within this limit. Multifunction instructions allow many common algorithms to be implemented within this limit.

- For cache contents to be valid, the cached instructions must be located in contiguous addresses in program memory. This means that DO UNTIL loops and JUMP loops may be used as long as the top-of-loop instruction is located within the sixteen-instruction region.

- Cache memory is only used as the instruction source when the PMD bus is needed for a data fetch that would conflict with the next instruction fetch.

3 Program Control

3.7.4   Cache Memory Example

Listing 3.1 shows a subroutine that implements a simple sum-of-products FIR filter. This example illustrates the execution speed advantages resulting from use of the cache. (This routine is discussed in further detail in Chapter 5, "Digital Filters," of *Digital Signal Processing Applications Using the ADSP-2100 Family*.)

```
.MODULE fir_sub;

{   Single-Precision FIR Transversal Filter Subroutine

    Calling Parameters
        I0 -> Oldest input data value in delay line
        L0 = Filter length (N)
        I4 -> Beginning of filter coefficient table
        L4 = Filter length (N)
        M1,M5 = 1
        CNTR = Filter length - 1 (N-1)
                        ⋮
    Return Values
        MR1 = Sum of products (rounded and saturated)
        I0 -> Oldest input data value in delay line
        I4 -> Beginning of filter coefficient table

    Altered Registers
        MX0,MY0,MR

    Computation Time
        N - 1 + 5 + 2 cycles

    Coefficients & data values assumed to be 1.15 format. }

.ENTRY fir;

fir: MR=0, MX0=DM(I0,M1), MY0=PM(I4,M5);
     DO sop UNTIL CE;
sop:    MR=MR+MX0*MY0(SS), MX0=DM(I0,M1), MY0=PM(I4,M5);
     MR=MR+MX0*MY0(RND);
     IF MV SAT MR;
     RTS;

.ENDMOD;
```

Listing 3.1 FIR Filter Subroutine

Program Control   3

As the subroutine executes, the following actions take place:

Instruction	Operation(s) Performed
`MR=0,MX0=DM(I0,M1),MY0=PM(I4,M5);`	Clear MR register, load X and Y input registers of MAC with DM data and PM data operands.
`DO sop UNTIL CE;`	Set up DO UNTIL loop.
`MR=MR+MX0*MY0(SS),MX0=DM(I0,M1),` `                MY0=PM(I4,M5);`	Perform multiply-accumulate, fetch two new operands. (Executes *N–1* times.)
`MR=MR+MX0*MY0(RND);`	Perform final multiply-accumulate, rounding result.
`IF MV SAT MR;`	Check for overflow and saturate if necessary.
`RTS;`	Return from subroutine.

Here is an overview of how the code actually executes in an ADSP-2100 system:

Cycle	Program Memory Bus	Processor Operation	Data Memory Bus
1	fetch 1st instruction	(execute previous instruction)	(previous instruction activity)
2	fetch operand 1	execute 1st instruction	fetch operand1
3	fetch 2nd instruction	(none—wait for 2nd instruction)	(no activity)
4	fetch 3rd instruction	execute 2nd instruction	(no activity)
5	fetch operand 2	execute 3rd instruction	fetch operand 2
6	fetch operand 3	execute 3rd instruction (from cache)	fetch operand 3
7	fetch operand 4	execute 3rd instruction (from cache)	fetch operand 4
8	fetch operand 5	execute 3rd instruction (from cache)	fetch operand 5
9	fetch 4th instruction	(none—wait for 4th instruction)	(no activity)
10	fetch 5th instruction	execute 4th instruction	(no activity)
11	fetch 6th instruction	execute 5th instruction	(no activity)
12	(fetch next instruction)	execute 6th instruction	(no activity)

The effect of the cache is most noticeable in cycles 5 through 8—the processor executes the multiply accumulate operation and fetches two operands in each cycle (the processor's data address generators also update the address pointers during each cycle). There is no penalty for fetching the operand from program memory.

Because the program sequencer allows zero overhead looping, the single instruction loop runs as fast as any straight-line coded version would; in addition, two busses are available for data fetches. This also results in compact code, in which the loop requires one set-up instruction and no overhead instructions for each iteration. An *N-tap* filter thus requires only $N+1$ cycles for the inner loop, while straight-line code might require $N * 2$ cycles with loop overhead included. Note that these ratios hold true for multiple instruction loops, not just for loops of one instruction.

Data Transfer ■ 4

4.1 OVERVIEW

This chapter describes the processor units that control the movement of data to and from the processor, and from one data bus to the other within the processor. These are the data address generators (DAGs) and the unit for exchanging data between the program memory data bus and the data memory data bus—the PMD-DMD bus exchange unit.

4.2 DATA ADDRESS GENERATORS (DAGS)

Every device in the ADSP-2100 family contains two independent data address generators so that both program and data memories can be accessed simultaneously. The DAGs provide indirect addressing capabilities. Both perform automatic address modification. For circular buffers, the DAGs can perform modulo address modification. The two DAGs differ: DAG1 generates only data memory addresses, but provides an optional bit-reversal capability, DAG2 can generate both data memory and program memory addresses, but has no bit-reversal capability.

While the following discussion explains the internal workings of the DAGs bear in mind that the ADSP-2100 family development software provides a direct method for declaring buffers as circular or linear and managing the placement of the buffer in memory. Only the initializing of DAG registers needs to be explicitly programmed. See the discussion of data structures in Chapter 12, "Instruction Set Reference."

4.2.1 DAG Architecture

Figure 4.1, which can be found on the following page, shows a block diagram of a single data address generator. There are three register files: the modify (M) register file, the index (I) register file, and the length (L) register file. Each of the register files contains four 14-bit registers which can be read from and written to via the DMD bus.

The I registers (I0-3 in DAG1, I4-7 in DAG2) contain the actual addresses used to access memory. When data is accessed in indirect mode, the address stored in the selected I register becomes the memory address.

4 Data Transfer

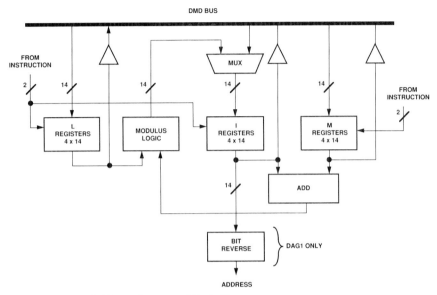

Figure 4.1 Data Address Generator Block Diagram

With DAG1, the output address can be bit-reversed by setting the appropriate mode bit in the mode status register (MSTAT) as discussed below or by using the ENA BIT_REV instruction. Bit-reversal facilitates FFT addressing.

The data address generators employ a post-modify scheme; after an indirect data access, the specified M register (M0-3 in DAG1, M4-7 in DAG2) is added to the specified I register to generate the updated I value. The choice of the I and M registers are independent within each DAG. In other words, any register in the I0-3 set may be modified by any register in the M0-3 set in any combination, but not by those in DAG2 (M4-7). The modification values stored in M registers are signed numbers so that the next address can be either higher or lower.

The address generators support both linear addressing and circular addressing. The value of the L register corresponding to an I register (for example, L0 would correspond to I0) determines which addressing scheme is used for that I register. For circular buffer addressing, the L register is initialized with length of the buffer. For linear addressing, the modulus logic is disabled by setting the corresponding L register to zero.

Data Transfer   4

Each time an I register is selected, the corresponding L register provides the modulus logic with the length information. If the sum of the M register and the I register crosses the buffer boundary, the modified I register value is calculated by the modulus logic using the L register value.

All data address generator registers (I, M, and L registers) are loadable and readable from the lower 14 bits of the DMD bus. Since I and L register contents are considered to be unsigned, the upper 2 bits of the DMD bus are padded with zeros when reading them. M register contents are signed; when reading an M register, the upper 2 bits of the DMD bus are sign-extended.

4.2.2   Modulo Addressing

The modulus logic implements automatic modulo addressing for accessing circular buffers. To calculate the next address, the modulus logic uses the following information:

- The current location, found in the I register (unsigned).
- The modify value, found in the M register (signed).
- The buffer length, found in the L register (unsigned).
- The buffer base address.

From these inputs, the next address is calculated according to the formula:

Next Address = (I + M - B) Modulo (L) + B

where:

I	=	current address
M	=	modify value (signed)
B	=	base address
L	=	buffer length
M+I	=	modified address

The inputs are subject to the condition:

$$|M| < L$$

This condition insures that the next address cannot wrap around the buffer more than once in one operation.

4   Data Transfer

4.2.3   Calculating The Base Address

In family processors other than the ADSP-2100, the base address of a circular buffer of length L is 2^n or a multiple of 2^n, where n satisfies the condition:

$$2^{n-1} < L \le 2^n$$

In other words, the base address is L "rounded" upwards to the closest power of 2 (or its multiple). In the ADSP-2100 processor, n must satisfy the condition:

$$2^{n-1} \le L < 2^n$$

The two rules differ only when L is itself a power of two. Both rules imply that a certain number of low-order bits of the base address must be zero; where L is a power of two, the ADSP-2100 requires one more such zero than the other processors in the family.

In practice, you do not need to calculate n yourself; the linker automatically places circular buffers at a proper address.

4.2.3.1 Circular Buffer Base Address Example 1

For example, let us assume that the buffer length is eight. According to the rule for processors other than the the ADSP-2100, the length of the buffer (eight) must be less than or equal to some value 2^n; n therefore, must be three or greater. The left side of the inequality specifies that the buffer length must be greater than the value 2^{n-1}; n therefore must be three or less. The only value of n that satisfies both inequalities is three. Valid base addresses are multiples of 2^n, so in this example valid base addresses are multiples of eight: 0x0008, 0x0010, 0x0018, and so on.

According to the rule for the ADSP-2100 processor, however, n must equal four; the valid base addresses are therefore multiples of sixteen: 0x0010, 0x0020, and so on.

4.2.3.2 Circular Buffer Base Address Example 2

As a second example, assume a buffer length of seven. Both rules yield the same value for n, namely, three. With a buffer length of seven, therefore, the valid base addresses are, in all family processors, multiples of eight: 0x0008, 0x0010, 0x0018, and so on.

Data Transfer   4

4.2.3.3 Circular Buffer Operation Example 1
Suppose that I0 = 5, M0 = 1, L0 = 3, and the base address = 4. The next
address is calculated as:

(I0 + M0 - B) mod L0 +B = (5 + 1 - 4) mod 3 + 4 = 6

The successive address calculations using I0 for indirect addressing
produce the sequence: 5, 6, 4, 5, 6, 4, 5…. For M0 = –1 (0x3FFF), I0 would
produce the sequence: 5, 4, 6, 5, 4, 6, 5, 4….

4.2.3.4 Circular Buffer Operation Example 2
Assume that I0 = 9, M0 = 3, L0 = 5, and the base address = 8. The 5-word
buffer resides at locations 8 through 12 inclusive. The next address is
calculated as:

(I0 + M0 - B) mod L0 +B = (9 + 3 - 8) mod 5 + 8 = 12

The successive address calculations using I0 for indirect addressing
produce the sequence: 9, 12, 10, 8, 11, 9... This example highlights the fact
that the address sequence does not have to result in a "direct hit" of the
buffer boundary.

4.2.4   Bit-Reverse Addressing
The bit-reverse logic is primarily intended for use in FFT computations
where inputs are supplied or the outputs generated in bit-reversed order.
Bit-reversing is available only on addresses generated by DAG1. The pivot
point for the reversal is the midpoint of the 14-bit address, between bits 6
and 7. This is illustrated in the following chart.

Individual DMA lines (DMA_N)

Normal Order	13	12	11	10	09	08	07	06	05	04	03	02	01	00
Bit-reversed	00	01	02	03	04	05	06	07	08	09	10	11	12	13

Bit-reversed addressing is a mode, enabled and disabled by setting a
mode bit in the mode status register (MSTAT). When enabled, all
addresses generated using index registers I0-3 are bit-reversed upon
output. (The modified valued stored back after post-update remains in
normal order.) This mode continues until the status bit is reset.

4 Data Transfer

It is possible to bit-reverse address values less than 14 bits wide. You must determine the first address and also initialize the M register to be used with a value calculated to modify the I register bit-reversed output to the desired range. This value is:

$$2^{(14-N)}$$

where N is the number of bits you wish to output reversed. *Digital Signal Processing Using the ADSP-2100 Family* also has a complete example of this in the chapter on Fast Fourier Transforms.

4.3   PMD-DMD BUS EXCHANGE

The PMD-DMD bus exchange unit couples the program memory data bus and the data memory data bus, allowing them to transfer data between them in both directions. Since the program memory data (PMD) bus is 24 bits wide, while the data memory data (DMD) bus is 16 bits wide, only the upper 16 bits of PMD can be directly transferred. An internal register (PX) is loaded with (or supplies) the additional 8 bits. This register can be directly loaded or read when the full 24 bits are required.

Note that when reading data from program memory and data memory simultaneously, there is a dedicated path from the upper 16 bits of the PMD bus to the Y registers of the computational units. This read-only path does not use the bus exchange circuit; it is the path shown on the individual computational unit block diagrams.

4.3.1   PMD-DMD Block Diagram Discussion

Figure 4.2 shows a block diagram of the PMD-DMD bus exchange. There are two types of connections provided in this section.

The first type of connection is a one-way path from each bus to the other. This is implemented with two tristate buffers connecting the DMD bus with the upper 16 bits of the PMD bus. One of these two buffers is normally used when data is exchanged between the program memory and one of the registers connected to the DMD bus. This is the path used to write data to program memory; it is not shown in the individual computational unit block diagrams.

Data Transfer   4

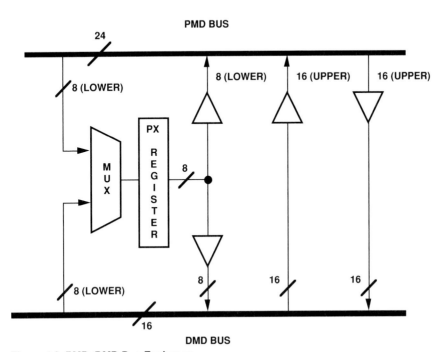

Figure 4.2 PMD–DMD Bus Exchange

The second connection is through the PX register. The PX register is 8-bits wide and can be loaded from either the lower 8 bits of the DMD bus or the lower 8 bits of the PMD bus. Its contents can also be read to the lower 8 bits of either bus.

PX register access follows the principles described below.

From the PMD bus, the PX register is:

1. Loaded automatically whenever data (not an instruction) is read from program memory to any register. For example:

   AX0 = PM(I4,M4);

   In this example, the upper 16 bits of a 24-bit program memory word are loaded into AX0 and the lower 8 bits are automatically loaded into PX.

4　Data Transfer

2. Read out automatically as the lower 8 bits when data is written to program memory. For example:

   PM(I4,M4) = AX0;

   In this example, the 16 bits of AX0 are stored into the upper 16 bits of a 24-bit program memory word. The 8 bits of PX are automatically stored to the 8 lower bits of the memory word.

From the DMD bus, the PX register may be:

1. Loaded with a data move instruction, explicitly specifying the PX register as the destination. The lower 8 bits of the data value are used and the upper 8 are discarded.

   PX = AX0;

2. Read with a data move instruction, explicitly specifying the PX register as a source. The upper 8 bits of the value read from the register are all zeroes.

   AX0 = PX;

Whenever any register is written out to program memory, the source register supplies the upper 16 bits. The contents of the PX register are automatically added as the lower 8 bits. If these lower 8 bits of data to be transferred to program memory (through the PMD bus) are important, you should load the PX register from DMD bus before the program memory write operation.

Serial Ports ■ 5

5.1   OVERVIEW

Synchronous serial ports, or SPORTs, support a variety of serial data communications protocols and can provide a direct interconnection between processors in a multiprocessor system.

These ADSP-2100 family processors contain serial ports:

Processor	Number of Serial Ports
ADSP-2101	2
ADSP-2102	2
ADSP-2105	1
ADSP-2106	1
ADSP-2111	2
ADSP-21msp50	2

The serial ports, designated SPORT0 and SPORT1 on processors with more than one, have some differences that are described in this chapter. On processors with one serial port, SPORT1 functions are provided.

5.2   BASIC SPORT DESCRIPTION

Each SPORT has a five-pin interface:

Pin Name	Function
SCLK	Serial clock
RFS	Receive frame synchronization
TFS	Transmit frame synchronization
DR	Serial data receive
DT	Serial data transmit

Table 5.1 SPORT External Interface

A SPORT receives serial data on its DR input and transmits serial data on its DT output. It can receive and transmit simultaneously, for full duplex operation. The data bits are synchronous to the serial clock SCLK, which is

5   Serial Ports

an output if the processor generates this clock or an input if the clock is generated externally. Frame synchronization signals RFS and TFS are used to indicate the start of a serial data word or stream of serial words.

Figure 5.1, shows a simplified block diagram of a single SPORT. Data to be transmitted is written from an internal processor register to the SPORT's TX register via the DMD bus. This data is optionally compressed in hardware, then automatically transferred to the transmit shift register. The bits in the shift register are shifted out on the SPORT's DT pin, MSB first, synchronous to the serial clock. The receive portion of the SPORT accepts data from the DR pin, synchronous to the serial clock. When an entire word is received, the data is optionally expanded, then automatically transferred to the SPORT's RX register, where it is available to the processor.

The following is a list of SPORT characteristics. Many of the SPORT characteristics are configurable to allow flexibility in serial communication.

- Bidirectional: each SPORT has independent transmit and receive sections.

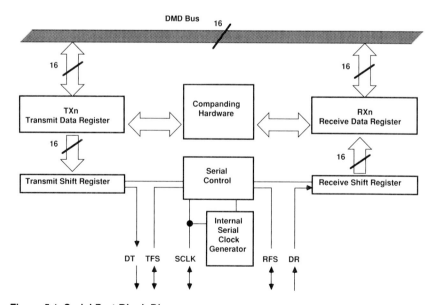

Figure 5.1 Serial Port Block Diagram

Serial Ports   5

- Double-buffered: each SPORT section (both receive and transmit) has a data register for transferring data words to and from other parts of the processor and a register for shifting data in or out. The double-buffering provides additional time to service the SPORT.

- Clocking: each SPORT can use an external serial clock or generate its own in a wide range of frequencies down to 0 Hz. See Section 5.5.

- Word length: each SPORT supports serial data word lengths from three to sixteen bits. See Section 5.6.

- Framing: each SPORT section (receive and transmit) can operate with or without frame synchronization signals for each data word; with internally-generated or externally-generated frame signals; with active high or active low frame signals; with either of two pulse widths and frame signal timing. See Section 5.7.

- Companding in hardware: each SPORT can perform A-law and μ-law companding according to CCITT recommendation G.711. See Section 5.10.

- Autobuffering with single-cycle overhead: using the DAGs, each SPORT can automatically receive and/or transmit an entire circular buffer of data with an overhead of only one cycle per data word. Transfers between the SPORT and the circular buffer are automatic in this mode and do not require additional programming. See Section 5.11.

- Interrupts: each SPORT section (receive and transmit) generates an interrupt upon completing a data word transfer, or after transferring an entire buffer if autobuffering is used. See Section 5.13.

- Multichannel capability: SPORT0 can receive and transmit data selectively from channels of a serial bitstream that is time-division multiplexed into 24 or 32 channels. This is especially useful for T1 interfaces or as a network communication scheme for multiple processors. See Section 5.12. **Note:** The ADSP-2105 has only one serial port (SPORT1) and does not support multichannel operation.

- Alternate configuration: SPORT1 can be configured as two external interrupt inputs, $\overline{IRQ0}$ and $\overline{IRQ1}$, and the Flag In and Flag Out signals instead of as a serial port. The internally generated serial clock may still be used in this configuration. See Section 5.4.

5 Serial Ports

5.2.1 Interrupts

Each SPORT has a receive interrupt and a transmit interrupt. The priority of these interrupts is shown in Table 5.2.

Highest   SPORT0 Transmit *(on 2-SPORT processors)*
           SPORT0 Receive *(on 2-SPORT processors)*
           SPORT1 Transmit
Lowest   SPORT1 Receive

Table 5.2 SPORT Interrupt Priorities

For complete details about how interrupts are handled, see the Interrupts section in Chapter 3, "Program Control."

5.2.2 SPORT Operation

Writing to a SPORT's TX register readies the SPORT for transmission; the TFS signal initiates the transmission of serial data. Once transmission has begun, each value written to the TX register is transferred to the internal transmit shift register and subsequently the bits are sent, MSB first. Each bit is shifted out on the rising edge of SCLK.

After the first bit (MSB) of a word has been transferred, the SPORT generates the transmit interrupt. The TX register is now available for the next data word, even though the transmission of the first word is ongoing.

In the receiving section, bits accumulate as they are received in an internal receive register. When a complete word has been received, it is written to the RX register and the receive interrupt for that SPORT is generated.

Interrupts are generated differently if autobuffering is enabled; see "Autobuffering" later in this chapter.

5.3 SPORT PROGRAMMING

To the programmer, the SPORT can be viewed as two sections. The configuration section is a block of control registers (mapped to data memory) that the program must initialize before using the SPORTs. The data section is a register file used to transmit and receive values through the SPORT.

Serial Ports   5

5.3.1   SPORT Configuration

SPORT configuration is accomplished by setting bit and field values in configuration registers. These registers are memory mapped in data memory space. SPORT0 configuration registers occupy locations 0x3FF3 to 0x3FFA; SPORT1 configuration registers occupy locations 0x3FEF to 0x3FF2. The contents of these registers are summarized in Table 5.3 and in the register summary in Appendix E. The effects of the various settings are described at length in the sections that follow.

Address	Contents
0x3FFA	SPORT0* multichannel receive word enables (31-16)
0x3FF9	SPORT0* multichannel receive word enables (15-0)
0x3FF8	SPORT0* multichannel transmit word enables (31-16)
0x3FF7	SPORT0* multichannel transmit word enables (15-0)
0x3FF6	SPORT0* control register
	Multichannel mode controls
	Serial clock source
	Frame synchronization controls
	Companding mode
	Serial word length
0x3FF5	SPORT0* serial clock divide modulus (determines frequency)
0x3FF4	SPORT0* receive frame sync divide modulus (determines frequency)
0x3FF3	SPORT0* autobuffer control register
0x3FF2	SPORT1 control register
	Flag output value
	Serial clock source
	Frame synchronization controls
	Companding mode
	Serial word length
0x3FF1	SPORT1 serial clock divide modulus (determines frequency)
0x3FF0	SPORT1 receive frame sync divide modulus (determines frequency)
0x3FEF	SPORT1 autobuffer control register (not on ADSP-21msp50/51)

*SPORT0 configuration registers are defined only on processors that have both SPORT0 and SPORT1

Table 5.3 SPORT Configuration Registers

There are two ways to initialize or to change values in SPORT configuration registers: write a register to an immediate address (instruction type 3) or write immediate data to an indirect address (instruction type 2). With either method, it is important to configure the serial port before enabling it.

5 Serial Ports

The first method of programming configuration registers requires no setup of DAG registers but does requires two instructions to perform the write. For example:

```
AX0 = 0x6B27;
DM(0x3FF2) = AX0;          {the contents of AX0 are written}
                           {to the address 0x3FF2}
AX0 = 0;
DM(0x3FF3) = AX0;          {the contents of AX0 are written}
                           {to address 0x3FF3}
```

In the second method, the DAG (I) index register must contain the data memory address of the configuration register to be written. The modify (M) register, which updates the I register after the write, must also contain a valid value. And the length (L) register that has the same number as the I register must be initialized to zero so that the circular buffer capability is not active. For example:

```
I0 = 0x3FF2;
M0 = 1;
L0 = 0;
DM(I0,M0) = 0x6B27;        {the constant 0x6B27 is written to }
                           {address pointed to by I0; pointer }
                           {then modified by M0}
DM(I0,M0) = 0;             {address 0x3FF3 is set to 0}
```

Either method works. The second method requires only one cycle to configure the registers once the I, M and L registers are initialized. This method is, however, more prone to error because the registers are written indirectly. You must make sure that the I register contains the intended value before the write.

5.3.2   Receiving And Transmitting Data

Each SPORT has a receive register and a transmit register. These registers are not memory mapped, but are identified by assembler mnemonics. The transmit registers are named TX0 and TX1, for SPORT0 and SPORT1 respectively. Receive registers are named RX0 and RX1 for SPORT0 and SPORT1 respectively. These registers can be accessed at any time during program execution using a data memory access with immediate address, load of a non-data register with immediate data or register-to-register move (instruction types 3, 7 and 17). For example, the following instruction would ready SPORT1 to transmit a serial value, assuming SPORT1 is configured and enabled:

```
TX1 = AX0;              {the contents of AX0 are transmitted}
                        {on SPORT1}
```

The following instruction would access a serial value received on SPORT0:

```
AYO = RX0;              {the contents of SPORT0 receive register}
                        {is transferred to AY0}
```

Because the SPORTs are interrupt driven, these instructions would typically be executed within a interrupt service routine in response to a SPORT interrupt.

5.4   SPORT ENABLE

SPORTs are enabled through bits in the system control register. This register is mapped to data memory address 0x3FFF. Bit 12 enables SPORT0 if it is a 1, and bit 11 enables SPORT1 if it is a 1. Both of these bits are cleared at reset, disabling both SPORTs.

Bit 10 of the system control register determines the configuration of SPORT1, either as a serial port or as interrupts and flags, according to Table 5.4 on the next page. If bit 10 is a 1, SPORT1 operates as a serial port; if it is a 0, the alternate functions are in effect (and bit 11 is ignored). At reset, bit 10 is a 1, so SPORT1 functions as a serial port.

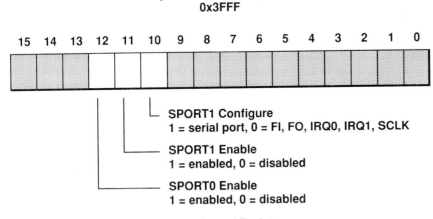

Figure 5.2 SPORT Enables In System Control Register

5   Serial Ports

Pin Name	Alternate Name	Alternate Function
RFS1	$\overline{IRQ0}$	External interrupt 0
TFS1	$\overline{IRQ1}$	External interrupt 1
DR1	FI	Flag input
DT1	FO	Flag output
SCLK1	Same	Same

Table 5.4 SPORT1 Alternate Configuration

5.5   SERIAL CLOCKS

Each SPORT operates on its own serial clock signal. The serial clock (SCLK) can be internally generated or received from an external source.

The ISCLK bit, bit 14 in either the SPORT0 or SPORT1 control register, determines the SCLK source for the SPORT. If this bit is a 1, the processor generates the SCLK signal; if it is a 0, the processor expects to receive an external clock signal on SCLK. At reset, ISCLK is cleared, so both serial ports are in the external clock mode. When ISCLK is set, internal generation of the SCLK signal begins on the next instruction cycle, whether or not the corresponding SPORT is enabled.

External serial clock frequencies may be as high as the processor's cycle rate up to 13 MHz; internal clock frequencies may be as high as one-half the processor's clock rate. The frequency of an internally generated clock is a function of the processor clock frequency (as seen at the CLKOUT pin) and the value of the 16-bit serial clock divide modulus register SCLKDIV (0x3FF5 for SPORT0 and 0x3FF1 for SPORT1).

SPORT0 Control Register: 0x3FF6
SPORT1 Control Register: 0x3FF2

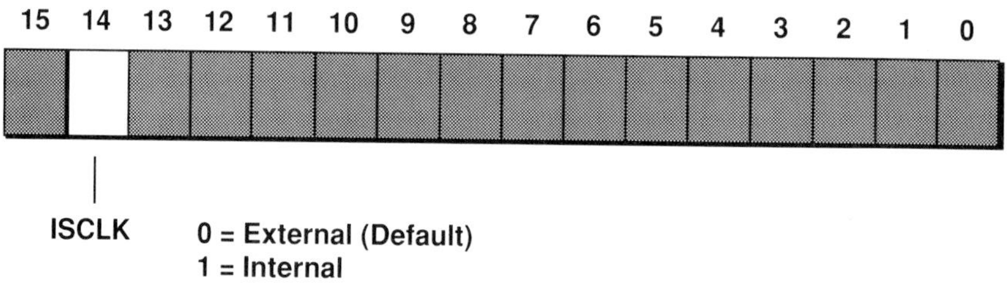

ISCLK   0 = External (Default)
          1 = Internal

Figure 5.3 ISCLK Bit In SPORT Control Register

96

Serial Ports   5

$$\text{SCLK frequency} = \frac{\text{CLKOUT frequency}}{2 \times (\text{SCLKDIV} + 1)}$$

Table 5.5 shows how some commonly required SCLK frequencies correspond to values of SCLKDIV.

SCLKDIV	SCLK Frequency
20479	300 Hz
5119	1200 Hz
639	9600 Hz
95	64 kHz
3	1.536 MHz
2	2.048 MHz
0	6.144 MHz

Assumes CLKOUT frequency of 12.288 MHz

Table 5.5 Common Internally Generated Serial Clock Frequencies

Note that the serial clock of SPORT1 (the SCLK pin) still functions when the port is being used in its alternate configuration (as FO, FI and two interrupts). In this case, SCLK is unresponsive to an external clock, but can internally generate a clock signal as described above.

5.6   WORD LENGTH

Each SPORT independently handles words of 3 to 16 bits. The data is right-justified in the SPORT data registers if it is fewer than 16 bits long. The serial word length (SLEN) field in each SPORT control register determines the word length according to this formula:

$$\text{Serial Word Length} = \text{SLEN} + 1$$

For example, if you are using 8-bit serial words, set SLEN to 7 (0111 binary). The SLEN field is bits 3-0 in the SPORT control register (0x3FF6 for SPORT0 and 0x3FF2 for SPORT1). See Figure 5.4 on the next page.

Do not set SLEN to zero or one; these SLEN values are not permitted.

5   Serial Ports

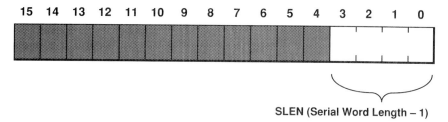

SPORT0 Control Register: 0x3FF6
SPORT1 Control Register: 0x3FF2

SLEN (Serial Word Length − 1)

Figure 5.4 SLEN Field In SPORT Control Register

5.7   WORD FRAMING OPTIONS

Framing signals identify the beginning of each serial word transfer. The
SPORTs have many ways of handling framing signals. Transmit and
receive framing are independent of each other.

5.7.1   Frame Synchronization

Word framing signals are optional. If the receive frame sync required
(RFSR) or transmit frame sync required (TFSR) bit in the SPORT control
register is a 0, a frame sync signal is necessary to initiate communications
but is ignored after the first bit is transferred. Words are then transferred

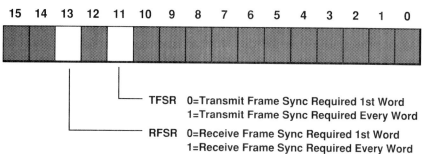

SPORT0 Control Register: 0x3FF6
SPORT1 Control Register: 0x3FF2

TFSR   0=Transmit Frame Sync Required 1st Word
       1=Transmit Frame Sync Required Every Word

RFSR   0=Receive Frame Sync Required 1st Word
       1=Receive Frame Sync Required Every Word

Figure 5.5 TFSR And RFSR Bits In SPORT Control Register

continuously, unframed. If the RFSR or TFSR bit is a 1, a frame sync signal is required at the start of every data word.

The RFSR bit is bit 13 in the SPORT control register (0x3FF6 for SPORT0 and 0x3FF2 for SPORT1), and the TFSR bit is bit 11. These bits are both cleared at reset, so that communication in both directions on both serial ports is unframed.

See "Configuration Examples" later in this chapter for examples of frame sync timing.

5.7.2   Frame Sync Signal Source

The processor can generate frame synchronization signals internally or receive them from an external source. The sources for transmit frame syncs and receive frames syncs can be set independently. If the internal receive frame sync (IRFS) bit or internal transmit frame sync (ITFS) bit in the SPORT control register is a 0, the processor expects to receive a signal on its frame sync pin (RFS or TFS). If the IRFS or ITFS bit is a 1, the processor generates its own frame sync signal and drives the RFS or TFS pin as an output.

The IRFS bit is bit 8 in the SPORT control register (0x3FF6 for SPORT0 and 0x3FF2 for SPORT1), and the ITFS bit is bit 9. Both of these bits are cleared at reset, that is, both serial ports require externally generated frame sync signals for both transmitting and receiving data.

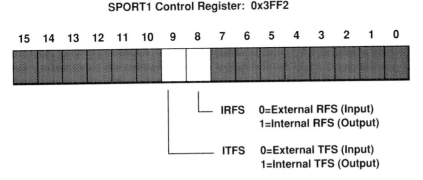

Figure 5.6 ITFS And IRFS Bits In SPORT Control Register

5 Serial Ports

If frame sync signals are generated externally, then RFS and TFS are inputs, and the external source controls data transmission and reception. The SPORT will wait for a transmit frame sync before transmitting data and for a receive frame sync before receiving data. If frame sync signals are generated internally, however, then RFS and TFS are outputs, and the processor controls the timing of data operations.

The SPORT outputs an internally generated transmit framing signal after data is loaded into the transmit (TX0 or TX1) register, at the time needed to ensure continuous data transmission, after the last bit of the current word is transmitted (the exact time depends on the framing mode being used; see "Normal and Alternate Framing Modes," the next section). The occurrence of the transmit frame sync is a result of the availability of data in the transmit register.

With an internally generated receive framing signal, the processor controls the timing of the receive data. The external data source must provide data to the serial port synchronized to the receive framing signal (the timing depends on the framing mode being used; see "Normal and Alternate Framing Modes," the next section). The processor generates RFS periodically on a multiple of SCLK cycles, based on the value of the 16-bit receive frame sync divide modulus register, RFSDIV (0x3FF4 for SPORT0 and 0x3FF0 for SPORT1):

Number of SCLK cycles between RFS assertions = RFSDIV + 1

For example, to allow 256 SCLK cycles between RFS assertions, set RFSDIV to 255 (0xFF).

Values of RFSDIV+1 that are less than the word length are not recommended.

Note that frame sync signals may be generated internally even when SCLK is supplied externally. This provides a way to divide external clocks for any purpose.

You can also use one frame sync to generate a single signal for both transmit and receive data. For example, an internally generated RFS (output) could be connected to an externally generated TFS (input) on the same SPORT for simultaneous transmit and receive operations. This interconnection is especially useful for combo codec interfaces.

Serial Ports 5

5.7.3 Normal And Alternate Framing Modes

In the normal framing mode, the framing signal is checked at the falling edge of SCLK. If the framing signal is asserted, received data is latched on the *next falling* edge of SCLK and transmitted data is driven on the *next rising* edge of SCLK. The framing signal is not checked again until the word has been transmitted or received. If data transmission or reception is continuous, i.e., the last bit of one word is followed without a break by the first bit of the next word, then the framing signal should occur in the same SCLK cycle as the last bit of each word.

In the alternate framing mode, the framing signal should be asserted in the same SCLK cycle as the first bit of a word. Received data bits are latched on the falling edge of SCLK and transmitted bits are driven on the rising edge of SCLK, but framing signal is checked only on the first bit. Internally generated frame sync signals remain asserted for the length of the serial word. Externally generated frame sync signals are only checked during the first bit time.

Framing modes for receiving and transmitting data are independent. If the receive frame sync width (RFSW) bit or transmit frame sync width (TFSW) bit in the SPORT control register is a 0, normal framing is enabled. If the RFSW or TFSW bit is a 1, alternate framing is used. The RFSW bit is bit 12 in the SPORT control register (0x3FF6 for SPORT0 and 0x3FF2 for SPORT1), and the TFSW bit is bit 10. These bits are both cleared at reset, so that normal framing in both directions is enabled.

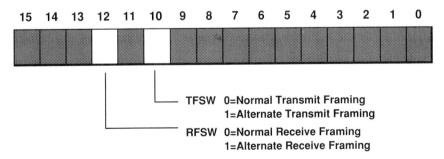

Figure 5.7 TFSW And RFSW Bits In SPORT Control Register

5 Serial Ports

For examples of normal and alternate framing, see "Configuration Examples" later in this chapter.

5.7.4 Active High Or Active Low

Framing sync signals for receiving and transmitting data can be either active high or active low and are configured independently. If the invert RFS (INVRFS) bit or invert TFS (INVTFS) bit in the SPORT control register is a 0, the corresponding frame sync signal is active high. If the INVRFS or INVTFS bit is a 1, the frame sync signal is active low. These controls apply regardless of the source of frame sync signals; they either control the polarity of internally generated signals or determine how externally generated signals are interpreted.

The INVRFS bit is bit 6 in the SPORT control register (0x3FF6 for SPORT0 and 0x3FF2 for SPORT1), and the INVTFS bit is bit 7. These bits are both cleared at reset, so that frame sync signals are active high.

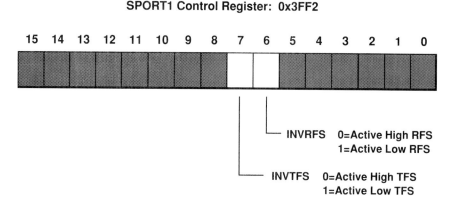

Figure 5.8 INVTFS And INVRFS Bits In SPORT Control Register

Serial Ports　5

5.8　CONFIGURATION EXAMPLE

The section of code that follows illustrates how to configure the SPORTs. This example configures both SPORT0 and SPORT1. SPORT0 is configured for an internally generated serial clock (SCLK), internally generated frame synchronization, and μ-law companded 8-bit data. This is a typical setup for communication with a combo codec. SPORT1 is configured for an externally generated serial clock, externally generated frame synchronization, non-companded 16-bit data and autobuffering. This setup could be used to transfer data between processors in a multiprocessor system.

Only the needed memory mapped registers are initialized. Notice that the SPORTs are configured before they are enabled and that any extraneous latched interrupts are cleared before interrupts are enabled.

```
{—    SPORT INITIALIZATION CODE    —}

{SPORT1 inits }

AX0 = 0x0017;
DM(0x3FEF) = AX0;        {enable SPORT1 autobuffering}
                        {TX autobuffer uses I0 and M0}
                        {RX autobuffer uses I1 and M1}

AX0 = 0x280F;
DM(0x3FF2) = AX0;        {external serial clock, RFS and TFS}
                        {RFS and TFS are required, normal}
                        {framing, no companding and 16 bits}

{SPORT0 inits}

{Assumes a CLKIN of 12.288 MHz. Internally generated}
{SCLK will be 2.048 MHz, and framing sync of 8 kHz}

AX0 = 255;
DM(0x3FF4) = AX0;        {RFSDIV = 256, 256 SCLKs between}
                        {frame syncs: 8 kHz framing}

AX0 = 2;
DM(0x3FF5) = AX0;        {SCLK = 2.048 MHz}
```

(continued on next page)

5 Serial Ports

(continued from previous page)

```
AX0 = 0x6B27;
DM(0x3FF6) = AX0;        {internal SCLK, RFS and TFS}
                        {normal framing, mu-law companding}
                        {8 bit words}


{SPORT ENABLE}

IFC = 0x1E;             {clear any extraneous SPORT interrupts}
ICNTL = 0;             {interrupt nesting disabled}

AX0 = 0x1C1F;          {both SPORTs enabled, BWAIT and}
DM(0x3FFF) = AX0;      {PWAIT left as default}

IMASK = 0x1E;          {SPORT interrupts are enabled}

{——   END SPORT INITIALIZATIONS   ——}
```

Figure 5.9 Example SPORT Configuration Code

5.9   TIMING EXAMPLES

This section contains examples of some combinations of the various framing options. The timing diagrams show relationships between signals, but are not scaled to show the actual timing parameters of the processor. Consult the data sheet for actual timing parameters and values.

The examples assume a word length of four bits, that is, SLEN = 3. Framing signals are active high, that is, INVRFS = 0 and INVTFS = 0.

The value of the SPORT control register (0x3FF6 for SPORT0 and 0x3FF2 for SPORT1) is shown for each example. In these binary values, 1= high, 0 = low, and X can be either. The **underlined** bit values are the bits which set the modes illustrated in the example.

Figures 5.10 to 5.15 show framing for receiving data. In Figures 5.10 and 5.11, the normal framing mode is shown for noncontinuous data (any number of SCLK cycles between words) and continuous data (no SCLK cycles between words). Figures 5.12 and 5.13 show noncontinuous and continuous receiving in the alternate framing mode. In these four figures, both the input timing requirement for an externally generated frame sync and the output timing characteristic of an internally generated frame sync are shown. Note that the output meets the input timing requirement; thus, on processors with two SPORTs, one SPORT could provide RFS for the other.

Serial Ports   5

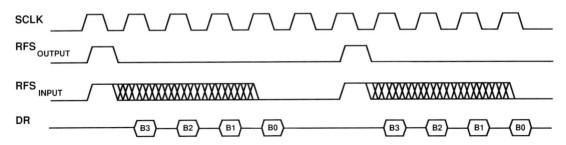

SPORT Control Register:
Internal Frame Sync   0X**10** XXX**1** X**0**XX 0011
External Frame Sync   0X**10** XXX**0** X**0**XX 0011
Both Internal Framing Option and External Framing Option Shown

Figure 5.10 SPORT Receive, Normal Framing

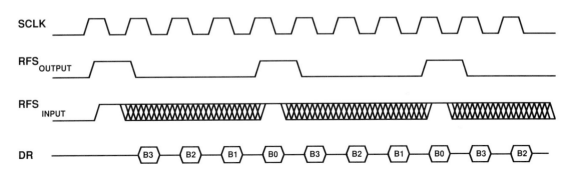

SPORT Control Register:
Internal Frame Sync   0X**10** XXX**1** X**0**XX 0011
External Frame Sync   0X**10** XXX**0** X**0**XX 0011
Both Internal Framing Option and External Framing Option Shown

Figure 5.11 SPORT Continuous Receive, Normal Framing

5 Serial Ports

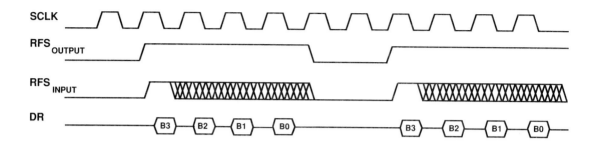

SPORT Control Register:
Internal Frame Sync   0X**11** XXX**1** X**0**XX 0011
External Frame Sync   0X**11** XXX**0** X**0**XX 0011
Both Internal Framing Option and External Framing Option Shown

Figure 5.12 SPORT Receive, Alternate Framing

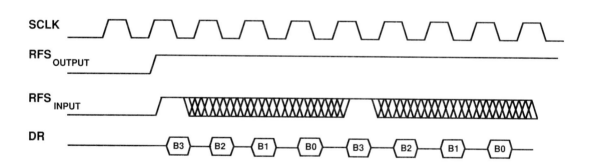

SPORT Control Register:
Internal Frame Sync   0X**11** XXX**1** X**0**XX 0011
External Frame Sync   0X**11** XXX**0** X**0**XX 0011
Both Internal Framing Option and External Framing Option Shown

Figure 5.13 SPORT Continuous Receive, Alternate Framing

Serial Ports   5

Figures 5.14 and 5.15 show the receive operation with normal framing and alternate framing, respectively, in the unframed mode. There is a single the frame sync signal that occurs only at the start of the first word, either one SCLK before the first bit (normal) or at the same time as the first bit (alternate). This mode is appropriate for multiword bursts (continuous reception).

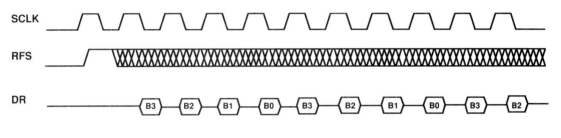

SPORT Control Register:
Internal Frame Sync   0X**00** XXX**1** X**0**XX 0011
External Frame Sync   0X**11** XXX**0** X**0**XX 0011

Figure 5.14 SPORT Receive, Unframed Mode, Normal Framing

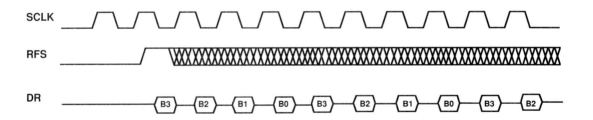

SPORT Control Register:
Internal Frame Sync   0X**01** XXX**1** X**0**XX 0011
External Frame Sync   0X**01** XXX**0** X**0**XX 0011

Figure 5.15 SPORT Receive, Unframed Mode, Alternate Framing

5 Serial Ports

Figures 5.16 to 5.21 show framing for transmitting data and are very similar to Figures 5.10 to 5.15. In Figures 5.16 and 5.17, the normal framing mode is shown for noncontinuous data and continuous data. Figures 5.18 and 5.19 show noncontinuous and continuous transmission in the alternate framing mode. As with receive timing, the TFS output meets the TFS input timing requirement.

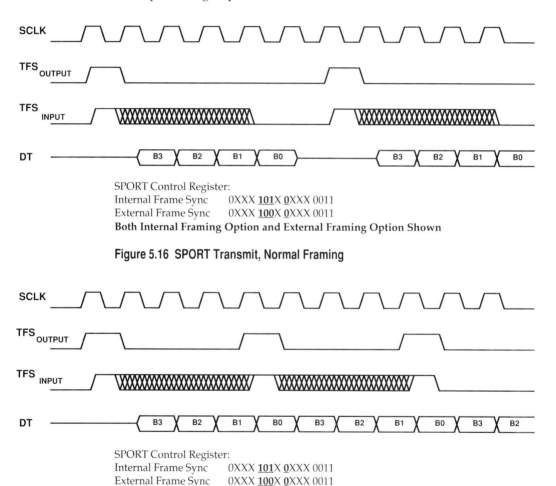

SPORT Control Register:
Internal Frame Sync   0XXX **101**X **0**XXX 0011
External Frame Sync   0XXX **100**X **0**XXX 0011
Both Internal Framing Option and External Framing Option Shown

Figure 5.16 SPORT Transmit, Normal Framing

SPORT Control Register:
Internal Frame Sync   0XXX **101**X **0**XXX 0011
External Frame Sync   0XXX **100**X **0**XXX 0011
Both Internal Framing Option and External Framing Option Shown

Figure 5.17 SPORT Continuous Transmit, Normal Framing

108

Serial Ports   5

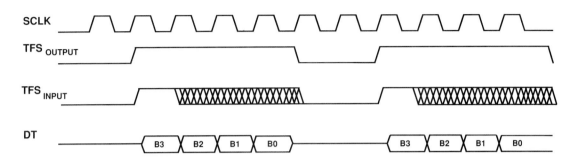

SPORT Control Register:
Internal Frame Sync   0XXX **111**X **0**XXX 0011
External Frame Sync   0XXX **110**X **0**XXX 0011
Both Internal Framing Option and External Framing Option Shown
Note: There is an asynchronous delay between TFS input and DT. See the appropriate data sheet for specifications.

Figure 5.18 SPORT Transmit, Alternate Framing

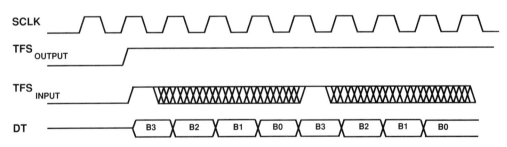

SPORT Control Register:
Internal Frame Sync   0XXX **111**X **0**XXX 0011
External Frame Sync   0XXX **110**X **0**XXX 0011
Both Internal Framing Option and External Framing Option Shown
Note: There is an asynchronous delay between TFS input and DT. See the appropriate data sheet for specifications.

Figure 5.19 SPORT Continuous Transmit, Alternate Framing

5   Serial Ports

Figures 5.20 and 5.21 show the transmit operation with normal framing and alternate framing, respectively, in the unframed mode. There is a single the frame sync signal that occurs only at the start of the first word, either one SCLK before the first bit (normal) or at the same time as the first bit (alternate).

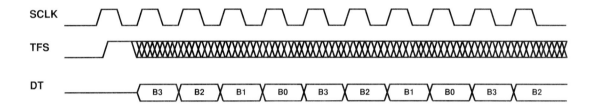

SPORT Control Register:
Internal Frame Sync   0XXX **001**X **0**XXX 0011
External Frame Sync   0XXX **000**X **0**XXX 0011

Figure 5.20 SPORT Transmit, Unframed Mode, Normal Framing

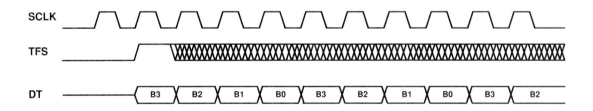

SPORT Control Register:
Internal Frame Sync   0XXX **011**X **0**XXX 0011
External Frame Sync   0XXX **010**X **0**XXX 0011
Note: There is an asynchronous delay between TFS input and DT. See the appropriate data sheet for specifications.

Figure 5.21 SPORT Transmit, Unframed Mode, Alternate Framing

110

5.10   COMPANDING AND DATA FORMAT

Companding (a contraction of COMpressing and exPANDing) is the process of logarithmically encoding and decoding data to minimize the number of bits that must be sent. Both SPORTs share the companding hardware; one expansion and one compression operation can occur in each processor cycle. In the event of contention, SPORT0 has priority.

The ADSP-2100 family of processors supports both of the widely used algorithms for companding: A-law and μ-law. The processor compands data according to the CCITT G.711 recommendation. The type of companding can be selected independently for each SPORT.

If companding is not enabled, there are two formats available for received data words of fewer than 16 bits: one that fills unused MSBs with zeros, and another that sign-extends the MSB into the unused bits.

The type of companding, as well as the non-companding data format, are controlled by the DTYPE field (bits 5-4) in the SPORT control register (0x3FF6 for SPORT0 and 0x3FF2 for SPORT1) as shown in Figure 5.22.

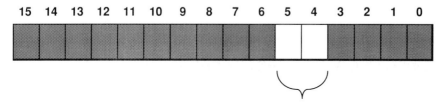

SPORT0 Control Register: 0x3FF6
SPORT1 Control Register: 0x3FF2

DTYPE   00=Right justify, zero fill unused MSBs
01=Right justify, sign extend into unused MSBs
10=Compand using μ-law
11=Compand using A-law

Figure 5.22 DTYPE Field In SPORT Control Register

5   Serial Ports

When companding is enabled, valid data in the RX0 or RX1 register is the right-justified, sign-extended, expanded value of the eight LSBs received. Likewise, a write to TX0 or TX1 causes the 16-bit value to be compressed to eight LSBs (sign-extended to the width of the transmit word) before being written to the internal transmit register. If the magnitude of the 16-bit value is greater than the 13-bit A-law or 14-bit μ-law maximum, the value is automatically compressed to the maximum positive or negative value.

5.10.1   Companding Operation Example

With hardware companding, interfacing to a codec requires little additional programming effort. See the codec hardware interfacing example in the last section of this chapter.

Here is a typical sequence of operations for transmitting companded data:

*   Write data to the TXn register
*   The value in TXn is compressed
*   The compressed value is written back to TXn
*   After the frame sync signal has occurred (if required), TXn is written to the internal transmit register and the bits are sent, MSB first.

As soon as the SPORT has started to send the second bit of the current word, TXn can be written with the next word, even though transmission of the first is not complete. After the MSB has been transferred, the SPORT generates the transmit interrupt to indicate that TXn is ready for the next data word. If the framing signal is being provided externally, the next word must be written to TXn early enough to allow for compression before the next framing signal arrives.

Here is a typical sequence of operations for receiving companded data:

*   Bits accumulate as received in the internal receive register
*   When a complete word is received, it is written to RXn
*   The value in RXn is expanded
*   The expanded value is written back to RXn

The receive interrupt for that SPORT is then generated.

Serial Ports   5

5.10.2   Contention For Companding Hardware

Since both SPORTs share the companding hardware, only one
compression and one expansion operation can take place during a single
machine cycle. If contention arises, such as when two expansions need to
occur in the same cycle, SPORT0 has priority, while SPORT1 is forced to
wait one cycle.

The effects of contention, however, are usually small. The instruction set
does not support loading both TX0 and TX1 in the same cycle;
consequently these operations will be naturally out of phase for
contention in many cases. The overhead cycle for the receive operation
occurs prior to the receive interrupt and does not increase the time needed
to service the interrupt, although it does affect the latency prior to
receiving the interrupt.

5.10.3   Companding Internal Data

Because the values in the RX and TX registers are actually companded "in
place" it is possible to use the companding hardware internally, without
any transmission or reception at all and without enabling the serial port.
This operation can be used for debugging or data conversion and requires
a single cycle of overhead.

To compress data, enable companding and then:

1. Write data to TXn (compression is calculated).
2. Wait for one cycle (TXn is written with compressed value)
3. Read TXn (it returns the 8-bit compressed data)

The code might look like this:

```
TX0 = AX0;      {linear data written to transmit register}
NOP;            {any instruction}
AX1 = TX0;      {compressed data transferred to AX1}
```

Use the same procedure to expand data, but use RXn instead of TXn.

```
RX0 = AX0;      {compressed data written to receive register}
NOP;            {any instruction}
AX1 = RX0;      {expanded - linear value transferred to AX1}
```

5 Serial Ports

5.11 AUTOBUFFERING

In normal operation, a SPORT generates an interrupt when it has received or has started to transmit a data word. Autobuffering provides a mechanism for receiving or transmitting an entire block of serial data before an interrupt is generated. Service routines can operate on the entire block of data, rather than on a single word, reducing overhead significantly. Autobuffering is available on both SPORT0 and SPORT1. except on the ADSP-21msp50/51 which autobuffers only on SPORT0.

Autobuffering uses the circular buffer addressing capability of the DAGs. With autobuffering enabled, each serial data word is transferred (or if multichannel operation is enabled, each active word is transferred) to or from data memory in a single overhead cycle. (Autobuffering to program memory is not supported.) This overhead cycle occurs independently of the instructions being executed and effectively suspends execution for one cycle (or more, if wait states are required) when it happens. No interrupt is generated for these individual data word transfers.

The autobuffer transfer cannot be duplicated by any instruction. However, an equivalent assembly language instruction would be:

```
DM(I,M) = RX0
    or                          Equivalent Instructions Only
TX0 = DM(I,M)
```

The I and M registers used in the transfer are selected by fields in the SPORT's autobuffer control register.

The processor waits for the current instruction to finish before inserting the overhead cycle. A delay in the autobuffer transfer occurs if the transfer is required during an instruction executing in multiple cycles (for wait states, for example). If the transfer is required when the processor is waiting in an IDLE state, the transfer is executed and the processor returns to IDLE.

When a data word transfer causes the circular buffer pointer to wrap around, the SPORT interrupt is generated. The receive interrupt occurs after the complete buffer has been received. The transmit interrupt occurs when the last word is loaded into TXn, prior to transmission.

Aside from the completion of an instruction requiring multiple cycles, the automatic transfer of individual data words has the highest priority of any operation short of RESET, including all interrupts. Thus, it is possible for an autobuffer transfer to increase the latency of an interrupt response if the interrupt happens to coincide with the transfer. Up to four autobuffered transfers can occur; in the case that two or more are needed in the same cycle, they have the following priority, which is the same as the SPORT interrupt priority:

Highest	SPORT0 Transmit
	SPORT0 Receive
	SPORT1 Transmit
Lowest	SPORT1 Receive

In the worst case that all four autobuffer transfers are required at about the same time, interrupt latency would increase by the time it takes for all the transfers to occur, which is affected by wait states and bus request.

5.11.1 Autobuffering Control Register

In autobuffering mode, an interrupt is generated when the modification of a specified I register (in the DAG) by the value in the specified M register (in the DAG) causes a modulus overflow (pointer wraparound). This means that the end of the buffer has been detected.

The autobuffering mode is enabled separately for receiving and transmitting by bits in the SPORT's autobuffer control register (0x3FF3 for SPORT0 or 0x3FEF for SPORT1), shown in Figure 5.23.

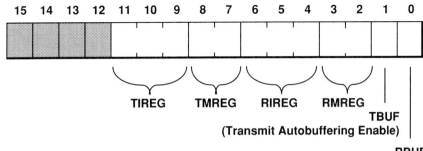

SPORT0 Autobuffer Control Register: 0x3FF3
SPORT1 Autobuffer Control Register: 0x3FEF

Figure 5.23 SPORT Autobuffer Control Register

115

5 Serial Ports

The I and M registers used for autobuffering are identified by fields in the autobuffer control register. TIREG and TMREG are binary values that indicate the numbers of the I and M registers, respectively, associated with the transmit buffer. The rules governing the pairing of I and M registers are the same as for other DAG operations: the I and M registers must be in the same DAG, numbered either 0-3 for DAG1 or 4-7 for DAG2. Consequently, three bits identify the I register, but only two bits are necessary to indicate the M register because the third bit (MSB) of the M register number must be the same as for the I register.

Likewise, RIREG and RMREG indicate the numbers of the I and M registers, respectively, associated with the receive buffer.

The TBUF and RBUF bits enable transmit autobuffering and receive autobuffering, respectively. These bits are cleared to zeros at reset and after a reboot. Consequently, autobuffering in progress cannot continue through a reboot operation; you must re-enable autobuffering after a reboot.

5.11.2 Autobuffering Example

The section of code that follows is an example that sets up SPORT1 for autobuffering operation. The code assumes that the processor is driven with a clock frequency of 12.288 MHz. The SPORT will automatically transmit values from the circular buffer named *tx_buffer*. It will receive values as they are sent to the SPORT and automatically transfer the data into the buffer named *rx_buffer*. A transmit interrupt will be generated once all of the *tx_buffer* values have been transferred to TX1, but before the last value has been loaded into the transmit shift register. A receive interrupt will be generated once the *rx_buffer* has been completely filled.

```
.MODULE/RAM    code_to_init_AB_SPORT1;

{— Initialization code for autobuffer —}

.VAR/DM/CIRC   tx_buffer[10];
.VAR/DM/CIRC   rx_buffer[10];
.ENTRY         sport1_inits;

{set up I,M, and L registers}
```

116

```
sport1_inits:   I0 = ^tx_buffer; {I0 contains address of tx_buffer}
                M0 = 1;           {fill every location}
                L0 = %tx_buffer; {L0 set to length of tx_buffer}

                I1 = ^rx_buffer; {I1 points to rx_buffer}
                L1 = %rx_buffer; {L1 set to length of rx_buffer}

{set up SPORT1 for autobuffering}

                AX0 = 0x0013;     {TX uses I0, M0; RX uses I1, M0}
                DM(0x3FEF) = AX0; {autobuffering enabled}

{set up SPORT1 for 8 kHz sampling and 2.048 MHz SCLK}

                AX0 = 255;        {set RFSDIV to 255 for 8 kHz}
                DM(0x3FF0) = AX0;

                AX0 = 2;      {set SCLKDIV to 2 for 2.048 MHz SCLK}
                DM(0x3FF5) = AX0;

{set up SPORT1 for normal required framing, internal SCLK}
{internal generated framing}

                AX0 = 0x6B27;     {normal framing, 8 bit mu-law}
                DM(0x3FF2) = AX0; {internal clock, framing}

{set up interrupts}

                IFC = 6;    {clear any extraneous SPORT interrupts}
                ICNTL = 0;  {interrupt nesting disabled}
                IMASK = 6;  {enable SPORT1 interrupts}

{enable SPORT1}

                AX0 = 0x0C1F;       {enable SPORT1 leave PWAIT,}
                DM(0x3FFF) = AX0;   {BWAIT as default}

{Place first transfer value into TX1}

                AX0 = DM(I0,M0);
                TX1 = AX0;
                RTS;
.ENDMOD;
```

Figure 5.24 Autobuffering Example Configuration Code

5   Serial Ports

5.12   MULTICHANNEL FUNCTION

SPORT0 supports a multichannel function. In the multichannel mode of operation, serial data is time-division multiplexed. Each subsequent word belongs to the next consecutive channel so that, for example, a 24-word block of data contains one word for each of 24 channels. SPORT0 supports 32 or 24 channels and can automatically select words for particular channels while ignoring the others.

In single-channel mode, receive and transmit framing identifies the start of a single word or continuous stream, with independent receive and transmit operation. In the multichannel mode, the receive frame sync signal (RFS0) identifies the start of a 24- or 32-word block of serial data with the receiver and transmitter operating in parallel. TFS0 has an alternate function, described below. **Note:** The ADSP-2105 has only one serial port (SPORT1) and does not support multichannel operation.

5.12.1   Multichannel Setup

Multichannel operation is enabled by bit 15 in SPORT0's control register (0x3FF6). When this bit is a 1, multichannel mode is enabled, and some control bits in the SPORT0 control register are redefined. Bits affected by multichannel mode are shown in Figure 5.25. At reset, bit 15 is cleared, disabling multichannel mode and enabling normal operation.

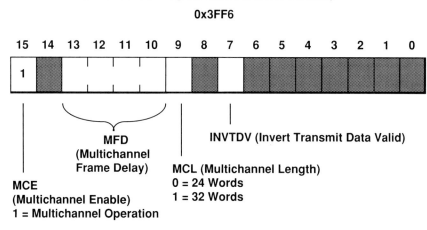

Figure 5.25 SPORT0 Control Register With Multichannel Mode Enabled

118

Serial Ports 5

The state of the multichannel length bit MCL, bit 9, determines whether there are 24 or 32 channels, i.e. whether the block length is 24 or 32 words. A 0 selects 24-word blocks; a 1, 32-word blocks. In multichannel mode, the word length is still set by the SLEN field in the SPORT control register and can be 3 to 16 bits.

The multichannel frame delay (MFD) is a 4-bit field specifying (in binary) the number of serial clock cycles between the frame sync signal and the first data bit. This allows the processor to work with different types of T1 interface devices. Figure 5.26 shows a variety of delays.

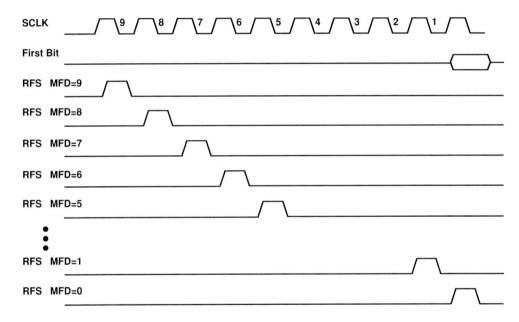

Figure 5.26 SPORT Multichannel Frame Delay Examples

The memory-mapped receive enable register and transmit enable register are each 32 bits wide and made up of two contiguous sixteen-bit registers, as shown in Figure 5.27, which can be found on the next page. Each bit corresponds to a channel; setting the bit enables that channel so that the processor will select its word from the 24- or 32-word block. For example, setting bit 0 selects word 0, bit 12 selects word 12, and so on.

5 Serial Ports

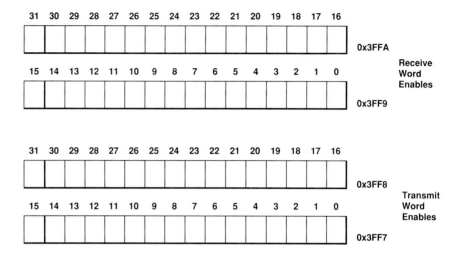

1 = Channel Enabled
0 = Channel Ignored

Figure 5.27 SPORT0 Multichannel Word Enable Registers

5.12.2 Multichannel Operation

Received words for channels that are not enabled are ignored; that is, no interrupts are generated for these words, no autobuffering occurs and no data is written to the RX0 register. Likewise, there are no interrupts and no autobuffering for transmit words that are not enabled. During transmit word time slots for channels that are not enabled, the data transmit (DT) pin is tristated.

Most aspects of SPORT0 operate normally in the multichannel mode. Specifically, word length (SLEN), internal or external framing (IRFS), frame signal inversion (INVRFS), companding (DTYPE) and autobuffering are unchanged in the multichannel mode. **Note:** It is important that RFS does not occur more than once per frame in multichannel mode.

Instead of providing frame synchronization, the TFS0 signal functions as a transmit data valid (TDV) signal in multichannel mode. TDV is asserted while the transmitter is active. TDV can be active high or low, and its polarity is controlled by the INVTFS bit, renamed INVTDV in this context. If INVTDV is a 1, TDV is active low; otherwise it is active high. TDV can be used to enable additional buffer logic, if required.

120

Serial Ports   5

Figure 5.28 shows the start of a multichannel transfer. As in earlier examples, word length is four bits (SLEN=3) and frame sync signals are active high. Multichannel frame delay (MFD) is one SCLK cycle. For the purpose of illustration, words 0 and 2 are selected for receiving and words 1 and 2 are selected for transmission.

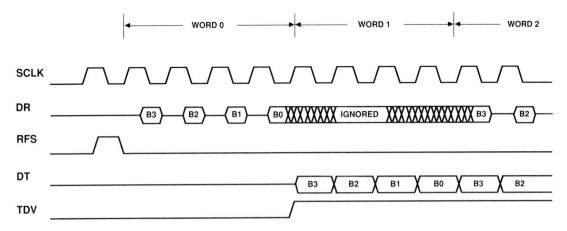

Figure 5.28 Start Of Multichannel Operation

Figure 5.29 (following page) shows a complete 24-word block in the multichannel mode, with complete words represented in the waveforms instead of individual bits. Receiving is active for all words and transmitting is active for words 0–3, 8–11 and 16–19 only.

Note: The ADSP-2105 has only one serial port (SPORT1) and does not support multichannel operation.

121

5 Serial Ports

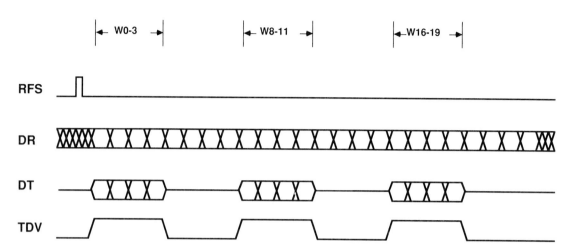

Figure 5.29 Complete Multichannel Example

5.13   SPORT TIMING CONSIDERATIONS

The SPORTs support full duplex operation and are normally interrupt driven. That is, whenever a SPORT transaction has completed, the processor generates an internal interrupt. Under most operating conditions, the actual timing of the SPORT interrupts is not critical. In some sophisticated DSP systems, however, it is important to know the timing of the interrupt relative to the operation of the serial port.

5.13.1   Companding Delay

Use of the companding circuit introduces latency in two ways. First, compressing or expanding a data value takes a single processor cycle. Second, SPORT0 has priority over SPORT1 if both require an expansion or compression operation in the same cycle; in this case, SPORT1 must wait one processor cycle. See the earlier section on companding in this chapter for more details on companding.

5.13.2   Clock Synchronization Delay

Some SPORT timings depend on the processor clock. Other timings depend on the serial clock (SCLK0 or SCLK1). These clocks are asynchronous. There is a delay associated with synchronizing the serial clock to the processor clock whether the serial clock is internally or externally generated. This delay is different for the transmit and receive interrupts, as explained in the following sections.

Serial Ports 5

5.13.3 Frame Sync Timing

In order to transmit data from a serial port all that is necessary, from the programmer's point of view, is to move the data into the appropriate TX register, using an instruction such as:

```
TX0 = AX0;
```

Once data is written into the TX register, a frame sync is generated after a synchronization delay. This latency in turn affects the timing of the serial port transmit interrupt. The latency depends on five factors: the frequency of the serial clock, whether or not companding is enabled, whether or not there is contention for the companding circuit, whether the current word has finished transmitting and the logic level of the SCLK when the data value was loaded into the transmit register.

If the transmit frame sync is generated externally, the data starts transmitting when the frame sync is received.

After the TX register is loaded, it takes three complete phases of the serial clock, HIGH, LOW and HIGH, in that order, to ensure synchronization (see Figure 5.30 on the next page.) Once synchronization has been ensured and a frame sync generated, the most significant bit of the transmit word is shifted out on the same rising edge as the frame sync if alternate framing is used and on the rising edge of the next serial clock if normal framing is used. Therefore, the worst-case synchronization delay is two SCLK cycles.

There is additional delay if the previous data transmission has not completed; the TX register cannot be loaded into the transmit shift register until the previous transmission is complete.

5.13.4 Transmit Interrupt Timing

Once the MSB has been transmitted, the subsequent bits are transmitted on the rising edges of the SCLK. The transmit interrupt (or autobuffer request) is generated internally on the falling edge of SCLK during the transmission of the second bit. This timing gives the program time to load the TX register with the next data for continuous data transmission.

The transmit interrupt, like any other interrupt, must be synchronized to the processor clock. Servicing is subject to the same latencies as other interrupts.

The transmit interrupt essentially means that it is all right to write a value to the TX register.

5 Serial Ports

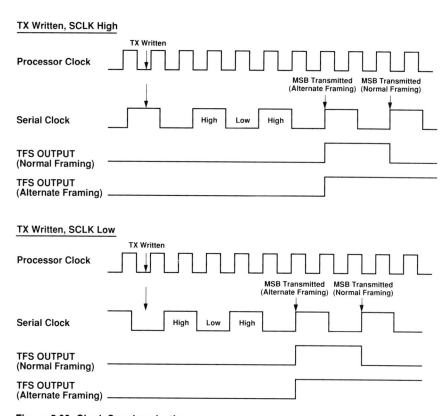

TX Written, SCLK High

Processor Clock

TX Written

MSB Transmitted
(Alternate Framing)　MSB Transmitted
(Normal Framing)

Serial Clock　　High　Low　High

TFS OUTPUT
(Normal Framing)

TFS OUTPUT
(Alternate Framing)

TX Written, SCLK Low

Processor Clock

TX Written

MSB Transmitted
(Alternate Framing)　MSB Transmitted
(Normal Framing)

Serial Clock　　High　Low　High

TFS OUTPUT
(Normal Framing)

TFS OUTPUT
(Alternate Framing)

Figure 5.30 Clock Synchronization

5.13.5   Receive Interrupt Timing

The receiver portion of the SPORT latches data on the DR pin on the falling edges of SCLK.

Receive interrupt timing differs from transmit interrupt timing. The receive interrupt or autobuffer request occurs only after an entire word is received. The interrupt request occurs on the rising edge of SCLK after a word is received and indicates that new data in the RX register can be read.

124

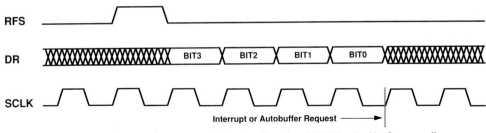

Figure 5.31 SPORT Interrupt Or Autobuffer Timing, Receive 4 Bit Words, No Companding

Companding causes a delay in the same manner as for transmitting. However, the latency is transparent, as the receive interrupt is generated after the expansion has taken place.

The LSB is received on the falling edge of SCLK. One processor cycle elapses to allow synchronization to the processor clock. One processor cycle later, the SPORT attempts to expand the data if companding is enabled and the other serial port is not using the companding circuitry. Companding latencies as discussed above occur prior to generation of a receive interrupt. Servicing the receive interrupt is subject to the same latencies as other interrupts.

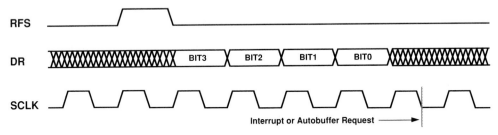

Figure 5.32 SPORT Interrupt Or Autobuffer Timing, Receive 4 Bit Words, Companding Enabled

5 Serial Ports

5.13.6   Startup Timing

The serial ports are treated as an asynchronous system to the processor even if the processor is providing the serial clock. Internal to the processor is a circuit which synchronizes the requests to the processor clock. The following diagram shows the synchronization delay for the serial ports, assuming the setup and hold times are met for the current processor cycle. The setup and hold times for the serial port requests are the same as shown on the data sheet for the $\overline{IRQ2}$ signal. If the setup and hold times are not met, there is an additional processor cycle of delay added.

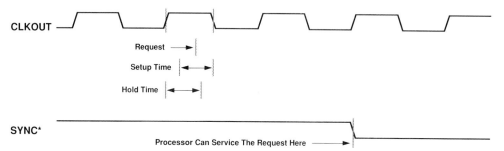

Figure 5.33 Synchronization Of Serial Clock To Processor Clock

As shown in the above diagram there is a two-processor cycle delay before the serial request is given to the processor. The same latencies exist for all external interrupts. The processor can only service interrupt or autobuffer requests on instruction cycle boundaries, so there may be additional latency cycles added due to the completion of an instruction.

5.13.7   Instruction Completion Latencies

There are several situations which can cause an instruction to take more than one processor cycle. Any of the following can delay the processor's ability to service a pending interrupt or autobuffer request:

*   External memory wait states
*   Bus request when an external access is required (in go-mode)
*   Bus request with go-mode disabled
*   Multiple external accesses required for a single instruction
*   A pending higher priority autobuffer or interrupt request
*   Interrupt being masked

On instruction cycle boundaries the processor will service multiple pending interrupt or autobuffer requests in the following priority order:

- SPORT0 transmit autobuffer (not on 2105) highest priority
- SPORT0 receive autobuffer (not on 2105)
- SPORT1 transmit autobuffer
- SPORT1 receive autobuffer
- Unmasked pending interrupts in priority order

5.13.8   Interrupt And Autobuffer Service Example

The following diagram shows the execution of a serial port interrupt based on a request that meets the setup and hold time requirements. This example is the same for a receive or a transmit interrupt request.

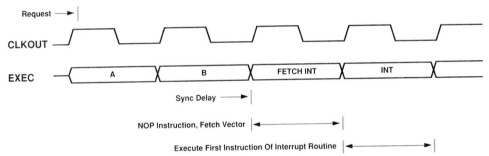

Figure 5.34 Interrupt Service Example

An additional latency cycle is consumed due to the fetching of the first instruction of the interrupt routine. The interrupt can only be serviced on an instruction cycle boundary. The above example assumes all instructions are completed in one processor cycle. The next diagram shows the result of an autobuffer request that meets the setup and hold requirements.

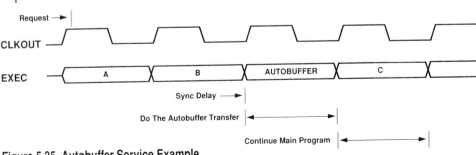

Figure 5.35 Autobuffer Service Example

5   Serial Ports

Autobuffering only consumes the cycles necessary to perform the data transfer; no additional cycles are lost fetching instructions. The above diagram assumes that all instructions and data transfers occur in one processor cycle.

5.13.9   Receive Companding Latency

In addition to the cycles used for synchronization, there are some additional delays possible due to receive companding. The synchronized request is used by the processor to decide when to write the receive register with the expanded value. This can only occur on instruction cycle boundaries and only one receive register can be expanded at a time. Since some of the ADSP-2100 family processors have more than one serial port, there is also a possibility of a delay due to the availability of the companding circuitry. SPORT0 has the higher priority. When companding is enabled, the autobuffer or interrupt request does not occur until the register has been expanded. The next two diagrams show examples of autobuffering with companding and the latencies involved.

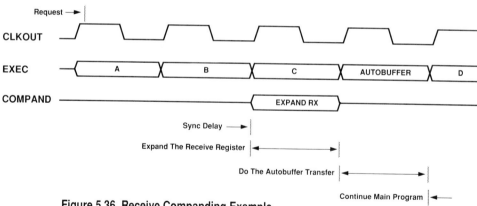

Figure 5.36 Receive Companding Example

Serial Ports   5

The following diagram shows the latency when there are two pending receive autobuffer requests with companding enabled.

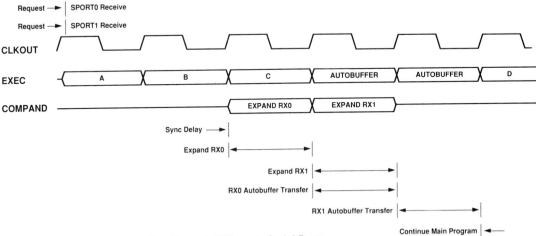

Figure 5.37 Receive Companding Example With Both Serial Ports

5.13.10 Interrupts With Autobuffering Enabled

When autobuffering is enabled, SPORT interrupts occur when the address modification done during the autobuffer operation causes a modulus wraparound. The synchronization delay applies to this type of interrupt as well. An example is shown in the next diagram.

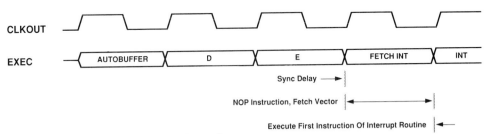

Figure 5.38 Autobuffering Interrupt Example

5   Serial Ports

5.13.11   Unusual Complications

In most cases the serial port companding, autobuffer, and interrupt latencies are transparent to the user's application program. When trying to use the same I register for more than one autobuffer channel, it becomes important to make sure that the latencies do not effect the correct order of operations. For example, if the serial port data is continuous, and the receiver and transmitter are working with the same frame signal, the order of the transmit and receive autobuffer or interrupt operations may be affected by the latencies shown above.

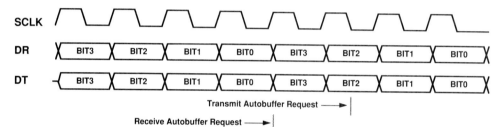

Figure 5.39 Using One Index Register For Transmit And Receive Autobuffer

If the processor is free to handle the autobuffer requests in the order they are generated, the receive autobuffer happens first and is then followed by the transmit autobuffer. The order of these operations may change if the processor is not available to handle the requests due to any of the previously mentioned latencies. In this case there are $1\frac{1}{2}$ serial clock cycles between the requests. If the processor is subject to bus requests, wait states, or other latencies which are longer than $1\frac{1}{2}$ serial clock cycles, both autobuffer operations may be held off. Since the transmit autobuffer has a higher priority, it's request will occur first. Because of the priority of the autobuffer requests the use of a single I register more difficult or even impossible in some cases. As long as there are no possible latency cases longer than the difference in the timing of the requests, it is quite possible to use a single I register for serial port autobuffering.

Timer ■ 6

6.1 OVERVIEW

The programmable interval timer can generate periodic interrupts based on multiples of the processor's cycle time. When enabled, a 16-bit count register is decremented every n cycles, where $n-1$ is a scaling value stored in an 8-bit register. When the value of the count register reaches zero, an interrupt is generated and the count register is reloaded from a 16-bit period register.

The scaling feature of the timer allows the 16-bit counter to generate periodic interrupts over a wide range of periods. Given a processor cycle time of 80 ns, the timer can generate interrupts with periods of 80 ns up to 5.24 ms with a zero scale value. When scaling is used, time periods can range up to 1.34 seconds.

Timer interrupts can be masked, cleared and forced in software if desired. For additional information, refer to the section "Interrupts" in Chapter 3, "Program Control."

6.2 TIMER ARCHITECTURE

The timer includes two 16-bit registers, TCOUNT and TPERIOD and one 8-bit register, TSCALE. The extended mode control instruction enables and disables the timer by setting and clearing bit 5 in the mode status register, MSTAT. For a description of the mode control instructions, refer to Chapter 12. The timer registers, which are memory-mapped, are shown in Figure 6.1 (on the following page).

TCOUNT is the count register. When the timer is enabled, it is decremented as often as once every instruction cycle. When the counter reaches zero, an interrupt is generated. TCOUNT is then reloaded from the TPERIOD register and the count begins again.

6 Timer

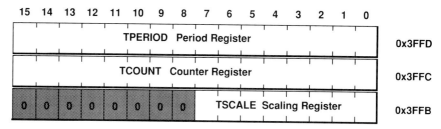

Figure 6.1 Timer Registers

TSCALE stores a scaling value that is one less than the number of cycles between decrements of TCOUNT. For example, if the value in TSCALE register is 0, the counter register decrements once every cycle. If the value in TSCALE is 1, the counter decrements once every 2 cycles. Figure 6.2 shows the timer block diagram.

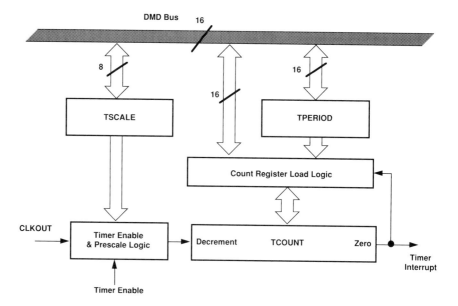

Figure 6.2 Timer Block Diagram

Timer 6

6.3 RESOLUTION

TSCALE provides the capability to program longer time intervals between interrupts, extending the range of the 16-bit TCOUNT register. Table 6.1 shows the range and the relationship between period length and resolution for TPERIOD = maximum.

Cycle Time = 80 ns

TSCALE	*Interrupt Every...*	*Resolution*
0	5.24 ms	80 ns
255	1.34 s	20 μs

Table 6.1 Timer Range And Resolution

6.4 TIMER OPERATION

Table 6.2 shows the effect of operating the timer with TPERIOD = 5, TSCALE = 1 and TCOUNT = 5. After the timer is enabled (cycle n–1) the counter begins. Because TSCALE is 1, TCOUNT is decremented on every other cycle. The reloading of TCOUNT and continuation of the counting occurs, as shown, during the interrupt service routine.

Cycle	*TCOUNT*	*Action*
n–4		TPERIOD loaded with 5
n–3		TSCALE loaded with 1
n–2		TCOUNT loaded with 5
n–1	5	ENA TIMER executed
n	5	since TSCALE = 1, no decrement
n+1	5	decrement TCOUNT
n+2	4	no decrement
n+3	4	decrement TCOUNT
n+4	3	no decrement
n+5	3	decrement TCOUNT
n+6	2	no decrement
n+7	2	decrement TCOUNT
n+8	1	no decrement
n+9	1	decrement TCOUNT
n+10	0	no decrement
n+11	0	zero reached, interrupt occurs load TCOUNT from TPERIOD
n+12	5	no decrement
n+13	5	decrement TCOUNT
n+14	4	no decrement
n+15	4	decrement TCOUNT, etc..

Table 6.2 Example Of Timer Operation

6 Timer

One interrupt occurs every (TPERIOD +1) * (TSCALE +1) cycles. To set the first interrupt at a different time interval from subsequent interrupts, load TCOUNT with a different value from TPERIOD. The formula for the first interrupt is (TCOUNT+1) * (TSCALE+1).

If you write a new value to TSCALE or TCOUNT, the change is effective immediately. If you write a new value to TPERIOD, the change does not take effect until after TCOUNT is reloaded.

Host Interface Port 7

7.1 OVERVIEW

The host interface port (HIP) is a parallel I/O port which allows some
ADSP-2100 family processors to act as memory-mapped peripherals (slave
DSPs) to a host computer. Examples of host computers include the ADSP-
2100 family, the Intel 8051 and the Motorola 68000 family. Table 1.1 in
Chapter 1 shows which family processors include the HIP.

The host interface port can be thought of as an area of dual-ported
memory, or mailbox registers, that allow communication between the host
and the processor core of the DSP. The host addresses the HIP as a section
of 8-bit or 16-bit words of memory. To the processor core, the HIP is a
group of eight data-memory-mapped registers.

Any number of ADSP-2100 family processors can be used in parallel as
memory-mapped peripherals. Assigning a different address location to
each DSP allows the host processor to control them all.

The speed of the HIP is similar to that of the processor data bus. A read or
write operation can occur within a single instruction cycle. Because the
HIP is normally connected with devices that are much slower (the 68000,
for example, can take four cycles to perform a bus operation), the data
transfer rate is usually limited by the host computer.

The host interface port is completely asynchronous to the rest of the DSP's
operations. The host can write data to or read data from the HIP while the
DSP is operating at full speed. The HIP can be configured for operation on
an 8-bit or 16-bit data bus and for either a multiplexed address/data bus
or separate address and data buses.

Those ADSP-2100 family processors which contain internal program RAM
and a HIP support two types of booting operations. One method boots
from external memory (usually EPROM) using the boot memory interface
described in Chapter 10. The other method uses the HIP to boot load a
program from a host computer. This second method is described later in
this chapter.

7 Host Interface Port

7.2 HIP PIN SUMMARY

The HIP consists of 27 pins. As shown in Table 7.1, 16 of these are data pins and 11 are control pins. Some of the control pins have dual functions, allowing the processor to support several bus protocols.

Pin Name	Number	Direction	Function
$\overline{\text{HSEL}}$	1	Input	HIP select
$\overline{\text{HACK}}$	1	Output	HIP acknowledge
HSIZE	1	Input	HIP 8/16 bit host 0=16-bit; 1=8-bit
BMODE	1	Input	HIP boot mode select 0=normal (2101); 1=HIP
HMD0	1	Input	HIP bus strobe select 0=$\overline{\text{RD}}$, $\overline{\text{WR}}$; 1=RW, $\overline{\text{DS}}$
$\overline{\text{HRD}}$/HRW[1]	1	Input	HIP **Read strobe**/ Read/Write select
$\overline{\text{HWR}}$/$\overline{\text{HDS}}$[1]	1	Input	HIP **Write strobe**/ Host data strobe
HMD1	1	Input	HIP address/data mode 0=separate; 1=multiplexed
HD15-0/HAD15-0[2]	16	Bidirectional	HIP **Data**/Address and data
HA2 /ALE[2]	1	Input	HIP **Host address 2**/ Address Latch Enable
HA1-0/no function[2]	2	Input	**Host addresses 1 and 0**
TOTAL	**27**		

[1] HMD0 selects function.
[2] HMD1 selects function.

Table 7.1 Host Interface Port Pins

136

Host Interface Port   7

$\overline{\text{HSEL}}$ is a host select which allows the host to enable or disable the HIP for host data transfers.

$\overline{\text{HACK}}$ is a host acknowledge output for hosts that require an acknowledge for handshaking.

HSIZE configures the bus size; the HIP can function in both 8-bit and 16-bit modes. If the HIP is configured for an 8-bit host (HSIZE=1), data is read from and written to the lower eight bits of a HIP data register and the upper eight bits are zero-filled (on host writes) or tristated (on host reads).

BMODE determines whether booting occurs through the HIP or through the memory interface pins.

HMD0 and HMD1 are mode pins that configure the address, data and strobe pins, as shown in Table 7.2. HMD0 configures the bus strobes, selecting either separate read and write strobes or a single read/write select and a host data strobe. HMD1 configures the bus protocol, selecting either separate address (3-bit) and data (16-bit) buses or a multiplexed 16-bit address/data bus with address latch enable. The timings of each of the four bus protocols are described later in this chapter.

	HMD1=0		HMD1=1	
HMD0=0	$\overline{\text{HRD}}$	HIP Read Strobe	$\overline{\text{HRD}}$	HIP Read Strobe
	$\overline{\text{HWR}}$	HIP Write Strobe	$\overline{\text{HWR}}$	HIP Write Strobe
	HD15-0	HIP Data	HAD15-0	HIP Address/Data
	HA2-0	HIP Address	ALE	HIP Address Latch Enable
HMD0=1	HRW	HIP Read/Write Select	HRW	HIP Read/Write Select
	$\overline{\text{HDS}}$	HIP Data Strobe	$\overline{\text{HDS}}$	HIP Data Strobe
	HD15-0	HIP Data	HAD15-0	HIP Address/Data
	HA2-0	HIP Address	ALE	HIP Address Latch Enable

Table 7.2 HIP Configuration Modes

137

7 Host Interface Port

The functions of the following pins are determined by HMD0 and HMD1 as described above:

HD15-0/HAD15-0 are either a data bus or a multiplexed address/data bus. (Only the 3 least significant address bits are used.)

$\overline{HRD}$/HRW is either a read strobe or a read/write select (1=read, 0=write).

$\overline{HWR}$/$\overline{HDS}$ is either a write strobe or a data strobe.

HA2/ALE is either the most significant host address bit or an address latch enable.

HA1-0 are either the two least significant host address bits or are unused.

7.3   HIP FUNCTIONAL DESCRIPTION

The HIP consists of three functional blocks, shown in Figure 7.1: a host control interface block (HCI), a block of six data registers (HDR5-0) and a block of two status registers (HSR7-6). The HIP also includes an associated HMASK register for masking interrupts generated by the HIP. The HCI provides the control for reading and writing the host registers. The two status registers provide status information to both the host and the DSP core.

The HIP data registers HDR5-0 are memory-mapped into internal data memory at locations 0x3FE0 (HDR0) to 0x3FE5 (HDR5). These registers can be thought of as a block of dual-ported memory. None of the HDRs are dedicated to either direction; they can be read or written by either the host or the DSP. When the host reads an HDR register, a maskable HIP read interrupt is generated. When the host writes an HDR, a maskable HIP write interrupt is generated.

The read/write status of the HDRs is also stored in the HSR registers. These status registers can be used to poll HDR status. Thus, data transfers through the HIP can be managed by using either interrupts or a polling scheme, described later in this chapter.

Host Interface Port   7

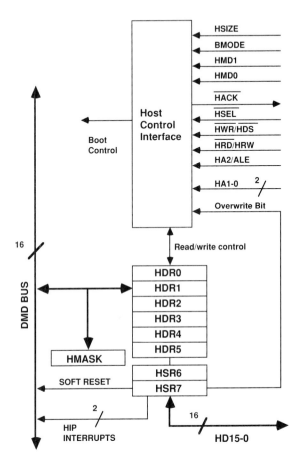

Figure 7.1 **HIP Block Diagram**

The HSR registers are shown in Figure 7.2, which can be found on the following page. Status information in HSR6 and HSR7 shows which HDRs have been written. The lower byte of HSR6 shows which HDRs have been written by the host computer. The upper byte of the HSR6 shows which HDRs have been written by the DSP. When an HDR register is read, the corresponding HSR bit is cleared.

7 Host Interface Port

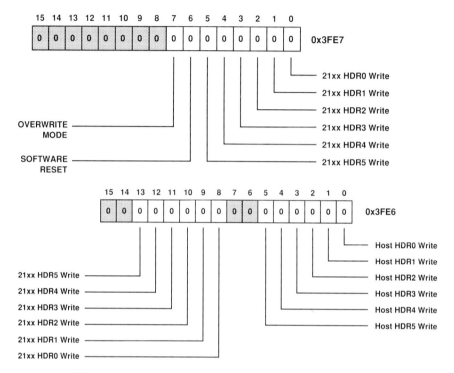

Figure 7.2 HIP Status Registers

The lower six bits of HSR7 are copied from the upper byte of HSR6 so that 8-bit hosts can read both sets of status. Bits 7 and 6 of HSR7 control the overwrite mode and software reset, respectively; these functions are described later in this chapter. The upper byte of HSR7 is reserved. All reserved bits and the software reset bit read as zeros. The overwrite bit is the only bit in the HSRs that can be both written and read. At reset, all HSR bits are zeros except for the overwrite bit, which is a one.

7.4   HIP OPERATION

The DSP core can place a data value into one of the HDRs for retrieval by the host computer. Similarly, the host computer can place a data value into one of the HDRs for retrieval by the DSP. To the host computer, the HDRs function as a section of memory. To the DSP, the HDRs are memory-mapped registers, part of the internal data memory space.

140

Host Interface Port   7

Because the HIP typically communicates with a host computer that has both a slower instruction rate and a multicycle bus cycle, the host computer is usually the limiting factor in the speed of HIP transfers. During a transfer, the DSP executes instructions normally, independent of HIP operation. This is true even during a multicycle transfer from the host.

For host computers that require handshaking, the DSP returns $\overline{\text{HACK}}$ in the same cycle as the host access, except in overwrite mode. In overwrite mode, the DSP can extend a host access by not asserting the $\overline{\text{HACK}}$ acknowledge until the cycle is complete. The user can enable and disable overwrite mode by setting and clearing a bit in HSR7. Overwrite mode is described in more detail later in this chapter.

The HDRs are not initialized during either hardware or software reset. The host can write information to the HDRs before a reset, and the DSP can read this information after the reset is finished. During reset, however, HIP transfers cannot occur; the $\overline{\text{HACK}}$ pin is de-asserted and the data pins are tristated.

Because a host computer that requires handshaking must wait for an acknowledgement from the DSP, it is possible to cause such a host to hang. If, when the host has initiated a transfer, but has not yet received an acknowledgement, the DSP is reset, then the acknowledgement can not be generated, thus causing the host to wait indefinitely.

There is no hardware in the HIP to prevent the host from writing a register that the DSP core is reading (or vice versa). If the host and the DSP try to write the same register at the same time, the host takes precedence. Simultaneous writes should be avoided, however: since the DSP and the host operate asynchronously, simultaneous writes can cause unpredictable results.

7.4.1   Polled Operation

Polling is one method of transferring data between the host and the DSP. Every time the host writes to an HDR, a bit is automatically set in the lower byte of HSR6. This bit remains set until the DSP reads the HDR. Similarly, when the DSP writes to an HDR, a bit in the upper byte of HSR6 (and the lower byte of HSR7) is set. This bit is cleared automatically when the host reads the HDR.

7 Host Interface Port

For example, the DSP can wait in a loop reading an HSR bit to see if the host has written new data. When the DSP sees that the bit is set, it conditionally jumps out of the loop, processes the new data, then returns to the loop. When transferring data to the host, the DSP waits for the host to read the last data written so that new data can be transferred. The host polls the HSR bits to see when the new data is available.

7.4.1.1 HIP Status Synchronization

Processes running on the DSP are asynchronous to processes running on the host. Values in shared registers can therefore change at any time, and reading a changing value can lead to unpredictable results. In one traditional approach to this problem of synchronization, the process receiving data reads a register twice and compares the results to see if the value changed. The overhead associated with the double read makes this an expensive solution. To eliminate the need for double reads, the ADSP-2100 family HIP has dual synchronizers which guarantee that the HIP status is constant during a read by either the processor core or the host. This synchronization is shown in Figures 7.3 and 7.4. A hardware or software handshaking protocol is recommended to avoid questionable data due to asynchronous operation.

The nature of host transfers is asynchronous with respect to the DSP processor clock. The status registers are updated by the DSP and are hence synchronous with the DSP processor clock. Synchronization circuitry built into the HIP insures status and data to be valid for either host or DSP access.

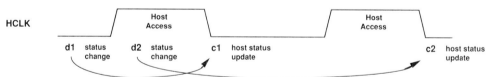

Figure 7.3 Host Status Synchronization

Host status synchronization is based on a pseudo clock HCLK, internal to the ADSP-21xx processor, which is a logical combination of $\overline{HRD}$, $\overline{HWR}$ and $\overline{HSEL}$. In Figure 7.3, a data change at d1 is shown. The status will be updated after the HCLK low, HCLK high, HCLK low sequence at point c1. A data change at d2 would wait for the HCLK low, HCLK high, HCLK low sequence and status would be updated at point c2.

Synchronization from the DSP requires one full CLKOUT cycle (starting at the rising edge) after a status change. As shown in Figure 7.4, a data change at point d1 would cause a status update at c1. A data change at d2 would cause a status update at c2.

142

Host Interface Port 7

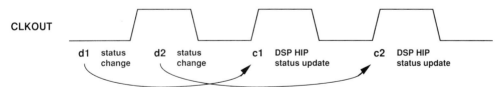

Figure 7.4 DSP HIP Status Synchronization

7.4.2   Interrupt-Driven Operation

Using an interrupt-driven protocol frees the host and the DSP from polling the HSR(s) to see when data is ready to be read. For interrupt-driven transfers to the DSP, the host writes data into an HDR, and the HIP automatically generates an internal interrupt. The interrupt is serviced like any other interrupt.

For transfers to the host, the DSP writes data to an HDR, then sets a flag output, which is connected to a host interrupt input, to signal the host that new data is ready to be transferred. Flag outputs are discussed in detail in Chapter 8. If the DSP passes data to the host through only one HDR, then that HDR can be read directly by the host when it receives the interrupt. If more than one HDR is used to pass data, then the host must read the appropriate HSR(s) to determine which HDR was written by the DSP.

7.4.3   HDR Overwrite Mode

In most cases, the DSP reads host data sent through the HIP faster than the host can send them. However, if the host is sufficiently fast, if the DSP is busy, or if the DSP is driven by a slow clock, there may be a delay in servicing a host write interrupt. If the host computer uses a handshaking protocol requiring the DSP to assert $\overline{\text{HACK}}$ to complete a host transfer, the DSP can optionally hold off the next host write until it has processed the current one.

If the HDR overwrite bit (bit 7 in HSR7) is cleared, and if the host tries to write to a register before it has been read by the DSP, $\overline{\text{HACK}}$ is not asserted until the DSP has read the previously written data. The host processor must wait for $\overline{\text{HACK}}$ to be asserted. As described earlier, however, there is a delay from when the host writes data to when the status is synchronized to the DSP. During this interval, it is possible for the host to write an HDR a second time even when the overwrite bit is cleared.

If the HDR overwrite bit is set, the previous value in the HDR is overwritten and $\overline{\text{HACK}}$ is returned immediately. If the DSP is reading the register that is being overwritten, the result is unpredictable.

7 Host Interface Port

On reset, the HDR overwrite bit is set. If the host does not require an acknowledge ($\overline{\text{HACK}}$ is not used), the HDR overwrite bit should be always be set, because there is no way for the DSP to prevent overwrite.

7.4.4 Software Reset

Writing a 1 to bit 6 of HSR7 causes software reset of the DSP. If the DSP writes the software reset bit, the reset happens immediately. Otherwise, the reset happens as soon as the write is synchronized to the DSP system clock. The internal software reset signal is held for five DSP clock cycles and then released.

7.5 HIP INTERRUPTS

HIP interrupts can be masked using either the IMASK register or the HMASK register. Bits in the IMASK register enable or disable all HIP read interrupts or all HIP write interrupts. The HMASK register, on the other hand, has bits for masking the generation of read and write interrupts for individual HDRs. In order for a read or write of an HDR to cause an interrupt, the HIP read or write interrupt must be enabled in IMASK, and the read or write to the particular HDR must be enabled in HMASK. HMASK is mapped to memory location 0x3FE8. IMASK is described in Chapter 3.

A host write interrupt is generated whenever the host completes a write to an HDR. A host read interrupt is generated when data written to an HDR has been read by the host, and the HDR is ready for another data word. The read interrupt is also active at reset, because no data has been read by the host yet; however, HMASK masks all HIP interrupts at reset. The read interrupt allows the DSP to transfer data to the host at a high rate without tying up the DSP with polling overhead.

HMASK allows reads and writes of some HDRs to not generate interrupts. For example, a system might use HDR2 and HDR1 for data values and HDR0 for a command value. Host write interrupts from HDR2 and HDR1 would be masked off, but the write interrupt from HDR0 would be unmasked, so that when the host wrote a command value, the DSP would process the command. In this way, the overhead of servicing interrupts when the host writes data values is avoided.

The HMASK register is organized in the same way as HSR6; the mask bit is in the same location as the status bit for the corresponding register. The lower byte of HMASK masks host write interrupts and the upper byte masks host read interrupts. The bits are all positive sense (0=masked, 1=enabled).

144

Host Interface Port 7

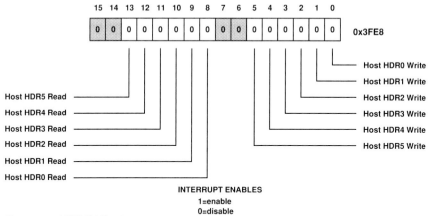

15 14 13 12 11 10 9 8 7 6 5 4 3 2 1 0

| 0 | 0 | 0 | 0 | 0 | 0 | 0 | 0 | 0 | 0 | 0 | 0 | 0 | 0 | 0 | 0 | 0x3FE8

Host HDR0 Write
Host HDR1 Write
Host HDR5 Read — Host HDR2 Write
Host HDR4 Read — Host HDR3 Write
Host HDR3 Read — Host HDR4 Write
Host HDR2 Read — Host HDR5 Write
Host HDR1 Read —
Host HDR0 Read —

INTERRUPT ENABLES
1=enable
0=disable

Figure 7.5 HMASK Register

HMASK is mapped to the internal data memory space at location 0x3FE8. On reset, the HMASK register is all zeros, which means that all HIP interrupts are masked.

HIP read and write interrupts are not cleared by servicing such an interrupt. Reading the HDR clears a write interrupt, and writing the HDR clears a read interrupt. The logical combination of all read and write interrupt requests generates a HIP interrupt. Pending interrupt requests remain until all HIP interrupts are cleared by either reading or writing the appropriate HIP data register. If the DSP is reading registers that the host might be writing, it is not certain that an interrupt will be generated. To ensure that all host writes generate interrupts, you must make sure that the DSP is not reading the HDRs that the host is writing. While servicing the interrupt, the status register can be read to determine which operation generated the interrupt and whether multiple interrupt requests need to be serviced.

HIP interrupts cannot be forced or cleared by software, as other interrupts can. The HIP write interrupt vector is location 0x0008. The HIP read interrupt vector is location 0x000C.

7.6   HOST INTERFACE TIMING

The following diagrams show the timings of HIP signals in the various modes determined by HMD0 and HMD1. HMD0 configures the bus strobes, selecting either separate read and write strobes or a single read/write select and a host data strobe. HMD1 configures the bus

7 Host Interface Port

protocol, selecting either separate address (3-bit) and data (16-bit) buses or a multiplexed 16-bit address/data bus with address latch enable. The HSIZE pin can be changed on a cycle-by-cycle basis; although not shown in the following diagrams, it has the same timing as the $\overline{\text{HRD}}$/HRW signal.

Figure 7.6 shows the HIP timing when both HMD0=0 and HMD1=0. HMD0 selects separate read and write strobes, and HMD1 selects separate address and data buses. The timing for the read cycle and the write cycle is as follows:

1. The host asserts the address.
2. The host asserts ($\overline{\text{HRD}}$ or $\overline{\text{HWR}}$) and $\overline{\text{HSEL}}$.
3. The DSP returns $\overline{\text{HACK}}$ (and, for a read cycle, the data).
4. For a write cycle, the host asserts the data.
5. The host de-asserts ($\overline{\text{HRD}}$ or $\overline{\text{HWR}}$) and $\overline{\text{HSEL}}$.
6. The host de-asserts the address (and, for a write cycle, the data).
7. The DSP de-asserts $\overline{\text{HACK}}$ (and, for a read cycle, the data).

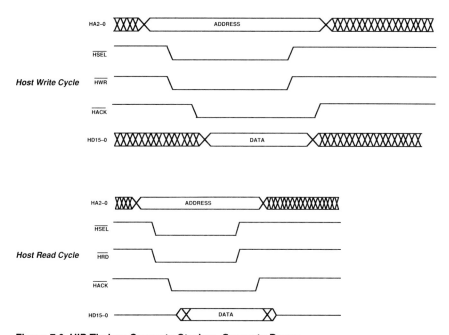

Figure 7.6 HIP Timing: Separate Strobes, Separate Buses

Host Interface Port   7

Figure 7.7 shows the HIP timing when HMD0=1 and HMD1=0. HMD0 selects a multiplexed read/write select with data strobe, and HMD1 selects separate address and data buses. The timing for the read cycle and the write cycle is as follows:

1. The host asserts HRW and the address.
2. The host asserts HDS and HSEL.
3. The DSP returns HACK (and, for a read cycle, the data).
4. For a write cycle, the host asserts the data.
5. The host de-asserts HDS and HSEL.
6. The host de-asserts HRW and the address (and, for a write cycle, the data).
7. The DSP de-asserts HACK (and, for a read cycle, the data).

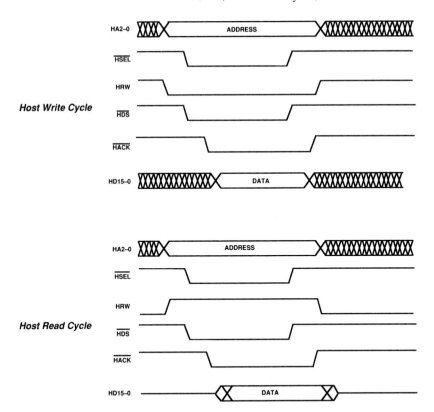

Figure 7.7 HIP Timing: Multiplexed R/W Strobe, Separate Buses

7 Host Interface Port

Figure 7.8 shows the HIP timing when HMD0=0 and HMD1=1. HMD0 selects separate read and write strobes, and HMD1 selects multiplexed address and data buses. HD0-HD2 are used for the address. The timing for the read cycle and the write cycle is as follows:

1. The host asserts ALE.
2. The host drives the address.
3. The host de-asserts ALE.
4. The host stops driving the address.
5. The host asserts (HRD or HWR) and HSEL.
6. The DSP returns HACK (and, for a read cycle, the data).
7. For a write cycle, the host asserts the data.
8. The host de-asserts (HRD or HWR) and HSEL.
9. For a write cycle, the host de-asserts the data.
10. The DSP de-asserts HACK (and, for a read cycle, the data).

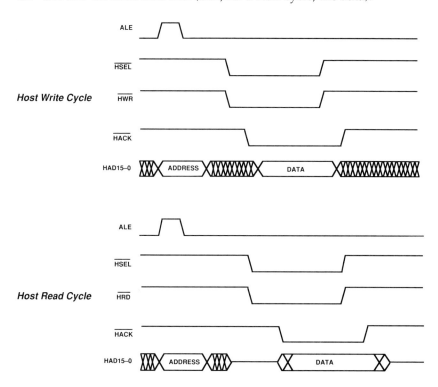

Figure 7.8 HIP Timing: Separate Strobes, Multiplexed Buses

148

Host Interface Port   7

Figure 7.9 shows the HIP timing when HMD0=1 and HMD1=1. HMD0 selects a multiplexed read/write select with data strobe, and HMD1 selects multiplexed address and data buses. HD0-HD2 are used for the address. The timing for the read cycle and the write cycle is as follows:

1. The host asserts ALE.
2. The host drives the address.
3. The host de-asserts ALE.
4. The host stops driving the address.
5. The host asserts HRW.
6. The host asserts $\overline{HDS}$ and $\overline{HSEL}$.
7. The DSP returns $\overline{HACK}$ (and, for a read cycle, the data).
8. For a write cycle, the host asserts the data.
9. The host de-asserts $\overline{HDS}$ and $\overline{HSEL}$.
10. The host de-asserts HRW (and, for a write cycle, the data).
11. The DSP de-asserts $\overline{HACK}$ (and, for a read cycle, the data).

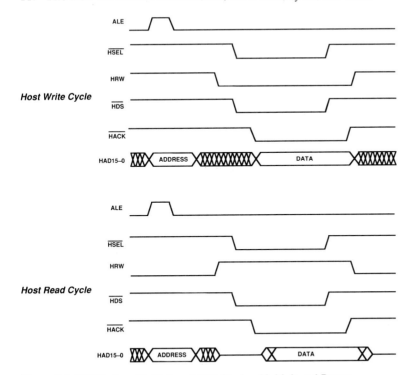

Figure 7.9 HIP Timing: Multiplexed R/W Strobe, Multiplexed Buses

7   Host Interface Port

7.7   BOOT LOADING THROUGH THE HIP

The entire internal program RAM of the DSP, or any portion of it, can be loaded using a boot sequence. Upon hardware or software reset, the boot sequence occurs if the MMAP pin is 0. If the MMAP pin is 1, the boot sequence does not occur.

The DSP can boot in either of two ways: from external memory (usually EPROM) using the boot memory interface, or from a host computer using the HIP. The BMODE pin specifies which type of booting occurs. When the BMODE pin is de-asserted, booting occurs through the memory interface. This process is described in Chapter 10. When the BMODE pin is asserted, booting occurs through the HIP. This process is described below.

The $\overline{\text{BMS}}$ (boot memory strobe) is asserted when booting through the HIP just as when booting through the memory interface; in this case, it serves as an indication that the boot sequence is occurring. Boot memory wait states have no effect when booting through the HIP.

Booting through the HIP occurs in the following sequence:

1. After reset, the host writes the length of the boot sequence to HDR3.

2. The host waits at least two DSP cycles.

3. Starting with the instruction which is to be loaded into the highest address of internal program memory, the host writes an instruction into HDR0, HDR2 and HDR1 (in that order), one byte each. The upper byte goes into HDR0, the lower byte goes into HDR2 and the middle byte goes into HDR1.

4. The address of the instruction is decremented, and Step 3 is repeated. This continues until the last instruction has been loaded into the HIP.

The DSP reads the length of the boot load first, then bytes are loaded from the highest address downwards. This results in shorter booting times for shorter loads.

Host Interface Port   7

The number of instructions booted must be a multiple of eight. The length value is given as:

length = (number of 24-bit program memory words/8) − 1

That is, a length of 0 causes the HIP to load eight 24-bit words.

In most cases, no handshaking is necessary, and the host can transfer data at the maximum rate that it is capable of. If the host operates faster than the DSP, wait states or NOPs must be added to the host cycle to slow it down to one write every DSP clock cycle.

The order in which the bytes are written (upper, lower, then middle) in Step 3 is the order in which the DSP transfers them into internal program memory. The DSP waits for a byte to be written to HDR0, then it reads HDR0, HDR2 and HDR1. If the host is writing bytes as fast as the DSP is reading them, then the host must always write in this order. If the host is slow enough, however, it can write the bytes in a different order, as long as the DSP can still transfer each byte from the HDR before the host writes a new one. In any case, each byte must always be written to the correct HDR.

A 16-bit host boots the DSP at the same rate as an 8-bit host. Either type of host must write the same data to the same the HDRs in the same sequence. If a 16-bit host writes 16-bit data, the upper byte of the data is loaded into the HDR but is not transferred to internal program memory in the booting sequence.

7 Host Interface Port

The following example shows the data that a host would write to the HIP for a 1000-instruction boot.

Data	*Location*
Page Length (124 decimal)	HDR3
Upper Byte of Instruction at 999	HDR0
Lower Byte of Instruction at 999	HDR2
Middle Byte of Instruction at 999	HDR1
Upper Byte of Instruction at 998	HDR0
Lower Byte of Instruction at 998	HDR2
Middle Byte of Instruction at 998	HDR1
Upper Byte of Instruction at 997	HDR0
Lower Byte of Instruction at 997	HDR2
Middle Byte of Instruction at 997	HDR1
•	•
•	•
•	•
Upper Byte of Instruction at 0	HDR0
Lower Byte of Instruction at 0	HDR2
Middle Byte of Instruction at 0	HDR1

Analog Interface ■ 8

8.1 OVERVIEW

The mixed-signal DSP processors of the ADSP-2100 family include an analog signal interface consisting of a 16-bit sigma-delta A/D converter, a 16-bit sigma-delta D/A converter, and a set of memory-mapped control and data registers. The analog interface offers the following features:

- linear-coded 16-bit sigma-delta ADC
- linear-coded 16-bit sigma-delta DAC
- on-chip anti-aliasing and anti-imaging filters
- 8 kHz sampling frequency
- programmable gain for DAC and ADC
- on-chip voltage reference

The analog interface provides a complete analog front end for high performance voiceband DSP applications. The ADC and DAC operate at a fixed sampling rate of 8 kHz. The inclusion of on-chip anti-aliasing and anti-imaging filters, 16-bit sigma-delta converters, and programmable gain amplifiers ensures a highly integrated solution to voiceband analog processing requirements. Sigma-delta conversion technology eliminates the need for complex off-chip anti-aliasing filters and sample-and-hold circuitry.

The ADSP-21msp50 is the first mixed-signal processor of the ADSP-2100 family. Other mixed-signal processors of the family, such as the ADSP-21msp55, contain the same analog interface as the ADSP-21msp50; they differ only in the amount of on-chip memory or in package and pin count. In this manual we refer only to the ADSP-21msp50 itself, but the description of the analog interface applies to all mixed-signal processors of the ADSP-2100 family; refer to the individual data sheet of each processor for on-chip memory and pinout specifics.

153

8 Analog Interface

The analog interface of the ADSP-21msp50 is operated with the use of several data-memory-mapped control and data registers. The ADC and DAC I/O can be transmitted and received via individual memory-mapped registers, or the data can be autobuffered directly into the processor's data memory. This autobuffering is similar to serial port autobuffering, as described in Chapter 5. Two ADSP-21msp50 interrupts are dedicated to the ADC and DAC converters. One interrupt is used for the ADC and the other interrupt is used for the DAC. Interrupts occur at the sample rate or when the autobuffer transfer is complete.

A block diagram of the analog interface is shown in Figure 8.1, and pin definitions are given in Table 8.1.

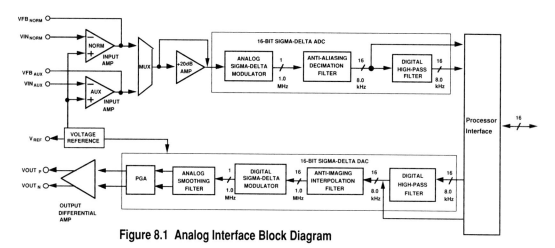

Figure 8.1 Analog Interface Block Diagram

Analog Interface   8

Pin Name	I/O	Function
VIN$_{NORM}$	I	Analog input to inverting terminal of NORM input amplifier.
VFB$_{NORM}$	O	Output terminal of NORM amplifier.
VIN$_{AUX}$	I	Analog input to inverting terminal of AUX input amplifier.
VFB$_{AUX}$	O	Output terminal of AUX amplifier.
VOUT$_P$	O	Analog output from non-inverting terminal of differential output amplifier.
VOUT$_N$	O	Analog output from inverting terminal of differential output amplifier.
V$_{REF}$	O	On-chip bandgap voltage reference (2.5 V ±10%).
V$_{CC}$	–	Analog supply voltage; nominal +5 V.
GND$_A$	–	Analog ground.

Table 8.1 Analog Interface Pin Definitions

8.2   A/D CONVERSION

The A/D conversion circuitry of the ADSP-21msp50's analog interface consists of two analog input amplifiers, an optional 20 dB pre-amplifier, and a sigma-delta analog-to-digital converter (ADC). The analog input signal must be ac-coupled.

8.2.1   Analog Input Amplifiers

The two analog input amplifiers (NORM, AUX) are internally biased by an on-chip voltage reference in order to allow operation of the ADSP-21msp50 with a single +5 V power supply.

An analog multiplexer selects either the NORM or AUX amplifier as the input to the ADC's sigma-delta modulator. The optional 20 dB pre-amplifier may be used to increase the signal level; the pre-amplifier can

8   Analog Interface

be inserted before the modulator or can be bypassed. Input signal level to the sigma-delta modulator should not exceed the V_{INMAX} specification listed in the *ADSP-21msp50 Data Sheet*. Refer to "Analog Input" in the "Design Considerations" section of this chapter for more information.

The input multiplexer and 20 dB pre-amplifier are configured by bits 0 and 1 (IPS, IMS) of the ADSP-21msp50's analog control register (which is memory-mapped at address 0x3FEE in data memory). The multiplexer setting should not be changed while an input signal is being processed.

8.2.2   ADC

The analog interface's ADC consists of a 2nd-order analog sigma-delta modulator, an anti-aliasing decimation filter, and a digital high pass filter. The sigma-delta modulator noise-shapes the signal and produces 1-bit samples at a 1.0 MHz rate. This bit stream, which represents the analog input signal, is fed to the anti-aliasing decimation filter.

8.2.2.1   Decimation Filter

The ADC's anti-aliasing decimation filter contains two stages. The first stage is a $sinc^4$ digital filter that increases resolution to 16 bits and reduces the sample rate to 40 kHz. The second stage is an IIR low pass filter.

The IIR low pass filter is a 10th-order elliptic filter with a passband edge at 3.7 kHz and a stopband attenuation of 65 dB at 4 kHz. This filter has the following specifications:

Filter type: 10th-order low pass elliptic IIR
Sample frequency: 40.0 kHz
Passband cutoff*:   3.70 kHz
Passband ripple: ±0.2 dB
Stopband cutoff: 4.0 kHz
Stopband ripple: −65.00 dB

* The passband cutoff frequency is defined to be the last point in the passband that meets the passband ripple specification.

(Note that these specifications apply only to this filter, and not to the entire ADC. The specifications can be used to perform further analysis of the exact characteristics of the filter, for example using a digital filter design software package.)

Figure 8.2 shows the frequency response of the IIR low pass filter.

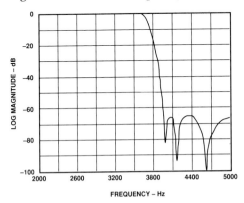

Figure 8.2 IIR Low Pass Filter Frequency Response

8.2.2.2   High Pass Filter

The ADC's digital high pass filter removes frequency components at the low end of the spectrum; it attenuates signal energy below the passband of the converter. The ADC's high pass filter can be bypassed by setting bit 7 (ADBY) of the ADSP-21msp50's analog control register.

The high pass filter is a 4th-order elliptic filter with a passband cutoff at 150 Hz. Stopband attenuation is 25 dB. This filter has the following specifications:

Filter type: 4th-order high pass elliptic IIR
Sample frequency: 8.0 kHz
Passband cutoff: 150.0 Hz
Passband ripple: ±0.2 dB
Stopband cutoff: 100.0 Hz
Stopband ripple: –25.00 dB

(Note that these specifications apply only to this filter, and not to the entire ADC. The specifications can be used to perform further analysis of the exact characteristics of the filter, for example using a digital filter design software package.)

8 Analog Interface

Figure 8.3 shows the frequency response of the high pass filter.

Passband ripple is ±0.2 dB for the combined effects of the ADC's digital filters (i.e. high pass filter and IIR low pass of the decimation filter) in the 300–3400 Hz passband.

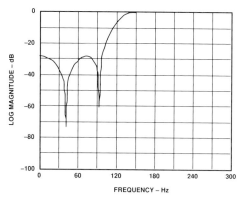

Figure 8.3 High Pass Filter Frequency Response

8.3   D/A CONVERSION

The D/A conversion circuitry of the ADSP-21msp50's analog interface consists of a sigma-delta digital-to-analog converter (DAC), an analog smoothing filter, a programmable gain amplifier, and a differential output amplifier.

8.3.1   DAC

The analog interface's DAC implements digital filters and a sigma-delta modulator with the same characteristics as the filters and modulator of the ADC. The DAC consists of a digital high pass filter, an anti-imaging interpolation filter, and a digital sigma-delta modulator.

The DAC receives 16-bit data values from the ADSP-21msp50's DAC Transmit data register (which is memory-mapped at address 0x3FEC in data memory). The data stream is filtered first by the DAC's high pass filter and then by the anti-imaging interpolation filter. These filters have the same characteristics as the ADC's anti-aliasing decimation filter and digital high pass filter.

Analog Interface   8

The output of the interpolation filter is fed to the DAC's digital sigma-delta modulator, which converts the 16-bit data to 1-bit samples at a 1.0 MHz rate. The modulator noise-shapes the signal such that errors inherent to the process are minimized in the passband of the converter.

The bit stream output of the sigma-delta modulator is fed to the DAC's analog smoothing filter where it is converted to an analog voltage.

8.3.1.1   High Pass Filter

The DAC's digital high pass filter has the same characteristics as the high pass filter of the ADC. The high pass filter removes frequency components at the low end of the spectrum; it attenuates signal energy below the passband of the converter. The DAC's high pass filter can be bypassed by setting bit 8 (DABY) of the ADSP-21msp50's analog control register.

The high pass filter is a 4th-order elliptic filter with a passband cutoff at 150 Hz. Stopband attenuation is 25 dB. This filter has the following specifications:

Filter type:	4th-order high pass elliptic IIR
Sample frequency:	8.0 kHz
Passband cutoff:	150.0 Hz
Passband ripple:	±0.2 dB
Stopband cutoff:	100.0 Hz
Stopband ripple:	−25.00 dB

(Note that these specifications apply only to this filter, and not to the entire DAC. The specifications can be used to perform further analysis of the exact characteristics of the filter, for example using a digital filter design software package.)

Figure 8.3 shows the frequency response of the high pass filter.

8.3.1.2   Interpolation Filter

The DAC's anti-imaging interpolation filter contains two stages. The first stage is is an IIR low pass filter that interpolates the data rate from 8 kHz to 40 kHz and removes images produced by the interpolation process. The output of this stage is then interpolated to 1.0 MHz and fed to the second stage, a $sinc^4$ digital filter that attenuates images produced by the 40 kHz to 1.0 MHz interpolation process.

8 Analog Interface

The IIR low pass filter is a 10th-order elliptic filter with a passband edge at 3.70 kHz and a stopband attenuation of 65 dB at 4 kHz. This filter has the following specifications:

Filter type: 10th-order low pass elliptic IIR
Sample frequency: 40.0 kHz
Passband cutoff*:   3.70 kHz
Passband ripple: ±0.2 dB
Stopband cutoff: 4.0 kHz
Stopband ripple: −65.00 dB

* The passband cutoff frequency is defined to be the last point in the passband that meets the passband ripple specification.

(Note that these specifications apply only to this filter, and not to the entire DAC. The specifications can be used to perform further analysis of the exact characteristics of the filter, for example using a digital filter design software package.)

Figure 8.2 shows the frequency response of the IIR low pass filter.

Passband ripple is ±0.2 dB for the combined effects of the DAC's digital filters (i.e. high pass filter and IIR low pass of the interpolation filter) in the 300–3400 Hz passband.

8.3.1.3 Analog Smoothing Filter & Programmable Gain Amplifier

The DAC's programmable gain amplifier (PGA) can be used to adjust the output signal level by −15 dB to +6 dB. This gain is selected by bits 2-4 (OG0, OG1, OG2) of the of the ADSP-21msp50's analog control register.

The DAC's analog smoothing filter consists of a 2nd-order Sallen-Key continuous-time filter and a 3rd-order switched capacitor filter. The Sallen-Key filter has a 3 dB point at approximately 80 kHz.

8.3.2 Differential Output Amplifier

The ADSP-21msp50's analog output signal ($VOUT_P - VOUT_N$) is produced by a differential amplifier. The differential amplifier can drive loads of 2 kΩ or greater and has a maximum differential output voltage swing of ±3.156 V peak-to-peak (3.17 dBm0). The output signal is dc-biased to the ADSP-21msp50's on-chip voltage reference (V_{REF})

Analog Interface   8

and can be ac-coupled directly to a load or dc-coupled to an external amplifier. Refer to "Analog Output" in the "Design Considerations" section of this chapter for more information.

The VOUTₚ – VOUTₙ outputs must be used as a differential signal; do not use either pin as a single-ended output.

8.4   OPERATING THE ANALOG INTERFACE

The analog interface of the ADSP-21msp50 is operated with the use of several memory-mapped control and data registers. The ADC and DAC I/O data can be received and transmitted in two memory-mapped data registers. The data can also be autobuffered into (and from) on-chip memory where data is automatically transferred to or from the data registers. In both cases, the I/O processing is interrupt-driven: two ADSP-21msp50 interrupts are dedicated to the analog interface, one for ADC receive data and one for DAC transmit data. **Note:** Autobuffering with SPORT1 is not available on the ADSP-21msp5x processors because this autobuffering channel is used for the analog interface.

The ADSP-21msp50 must have an input clock input frequency of 13 MHz. At this frequency, analog-to-digital and digital-to-analog converted data is transmitted at an 8 kHz rate with a single 16-bit word transmitted every 125 μs.

8.4.1   Memory-Mapped Control Registers

Two memory-mapped control registers are used to configure the ADSP-21msp50's analog interface: the analog control register and analog autobuffer/powerdown register.

8.4.1.1   Analog Control Register

The analog control register (located at address 0x3FEE in data memory) is shown in Figure 8.4. This register configures the ADC input multiplexer, ADC +20 dB preamplifier, ADC high pass filter, DAC output gain (PGA), and DAC high pass filter.

The analog control register also contains the APWD bits (bits 5, 6) which must both be set to ones to enable and start up the analog interface. The DAC and ADC begin transmitting data after these bits are set. Clearing the APWD bits disables the entire analog interface by putting it in a powerdown state. The APWD bits must be cleared (set to zeros) at least three processor cycles before putting the processor in powerdown. Refer to "Powerdown" in Chapter 9.

161

8 Analog Interface

The analog control register is cleared (set to 0x0000) by the processor's RESET signal. Note that bits 9-15 of this register are reserved and must always be set to zero.

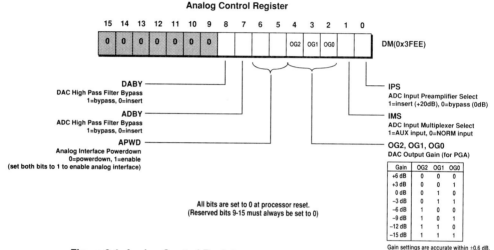

Figure 8.4 Analog Control Register

8.4.1.2 Analog Autobuffer/Powerdown Register

The analog autobuffer/powerdown register (located at address 0x3FEF in data memory) is shown in Figure 8.5. This register enables or disables autobuffering of ADC receive data and/or DAC transmit data—autobuffering is enabled by writing ones to the ARBUF (bit 0) and/or ATBUF (bit 1) bits. When autobuffering is enabled, I (index) and M (modify) registers are selected in bits 2–11 for the receive and/or transmit data buffers.

Bits 12–15 of the analog autobuffer/powerdown register control the ADSP-21msp50's processor powerdown function, *not* powerdown of the analog interface—powerdown of the analog interface only is controlled by the APWD bits (bits 5, 6) of the analog control register. The ADSP-21msp50's processor powerdown function is described in the next chapter, "System Interface."

Analog Interface   8

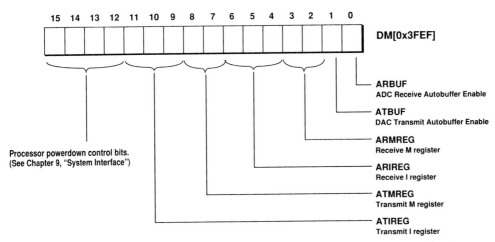

Analog Autobuffer/Powerdown Control Register

DM[0x3FEF]

ARBUF
ADC Receive Autobuffer Enable

ATBUF
DAC Transmit Autobuffer Enable

ARMREG
Receive M register

ARIREG
Receive I register

ATMREG
Transmit M register

ATIREG
Transmit I register

Processor powerdown control bits.
(See Chapter 9, "System Interface")

Figure 8.5 Analog Autobuffer/Powerdown Control Register (Autobuffer Control Bits Only)

8.4.2   Memory-Mapped Data Registers

There are two memory-mapped data registers dedicated to the analog interface. The 16-bit ADC receive data register is located at address 0x3FED in data memory. The 16-bit DAC transmit data register is located at address 0x3FEC in data memory. These registers must be individually read and written when autobuffering is not in use (autobuffering automatically transfers the data to and from processor data memory).

When autobuffering is disabled, data must be transmitted to the sigma-delta DAC by writing a 16-bit word to the DAC transmit register (0x3FEC) and data must be received from the sigma-delta ADC by reading a 16-bit word from the ADC receive register (0x3FED).

8.4.3   ADC & DAC Interrupts

The analog interface generates two interrupts that signal either: 1) that a 16-bit, 8 kHz analog-to-digital or digital-to-analog conversion has been completed, or 2) that an autobuffer block transfer has been completed (i.e. the data buffer is full or empty).

8 Analog Interface

When one of the analog interrupts occurs, the processor vectors to the appropriate address:

DAC Transmit interrupt vector address: 0x18
ADC Receive interrupt vector address:　0x1C

These interrupts can be masked out in the processor's IMASK register and can be forced or cleared in the IFC register.

8.4.3.1 Autobuffering Disabled

The ADC receive and DAC transmit interrupts occur at an 8 kHz rate, indicating when the data registers should be accessed, when autobuffering is disabled. On the receive side, the ADC interrupt is generated each time an A/D conversion cycle is completed and the 16-bit data word is available in the ADC receive register. On the transmit side, the DAC interrupt is generated each time a D/A conversion cycle is completed and the DAC transmit register is ready for the next 16-bit data word.

Both interrupts are generated simultaneously at an 8 kHz rate, occurring every 1625 instruction cycles with a 13.00 MHz processor clock, when autobuffering is disabled. The interrupts are generated continuously, starting when the analog interface is powered up by setting the APWD bits (bits 5, 6) to ones in the analog control register. Because both interrupts occur simultaneously, only one need be enabled (in IMASK) to vector to a single service routine which handles both transmit and receive data.

A simple analog loopback program is shown in Listing 8.1.

Analog Interface   8

```
{ ADSP-21msp50 Analog Interface Loopback Example      }
{   - configures analog interface                     }
{   - copies ADC receive data to DAC transmit buffer}


.MODULE/ABS=0/BOOT=0  talkthru;

#define  codec_tx_data   0x3FEC
#define  codec_rx_data   0x3FED
#define  codec_ctrl_reg  0x3FEE

resetv:         JUMP setup; NOP; NOP; NOP;
irq2v:          RTI; NOP; NOP; NOP;          {interrupt vectors ...}
hipwv:          RTI; NOP; NOP; NOP;
hiprv:          RTI; NOP; NOP; NOP;
spt0tv:         RTI; NOP; NOP; NOP;
spt0rv:         RTI; NOP; NOP; NOP;
antv:           RTI; NOP; NOP; NOP;
anrv:           SI = DM(codec_rx_data);      {read in data from ADC}
                DM(codec_tx_data) = SI;      {write out data to DAC}
                RTI; NOP;
irq1v:          RTI; NOP; NOP; NOP;
irq0v:          RTI; NOP; NOP; NOP;
timerv:         RTI; NOP; NOP; NOP;
pwrdwnv:        RTI; NOP; NOP; NOP;

setup:          AX1 = 0x0060;
                DM(codec_ctrl_reg) = AX1;  {power up analog interface}
                IMASK = 0x8;               {enable analog receive interrupt}
wait_loop:      IDLE;                       {wait for interrupt}
                JUMP wait_loop;
.ENDMOD;
```

Listing 8.1 ADSP-21msp50 Analog Loopback Program

8.4.3.2   Autobuffering Enabled

In some applications it is advantageous to perform block data transfers
between the analog converters and processor memory. Analog
interface autobuffering allows you to automatically transfer blocks of
data from the ADC to on-chip processor data memory or from on-chip
processor data memory to the DAC.

An interrupt is generated when an entire block transfer is complete (i.e.
when the data buffer is full or empty). Analog interface autobuffering
operates in the same way as SPORT autobuffering, described in
Chapter 5. Note that data can be autobuffered through the analog
converters or through SPORT0 of the ADSP-21msp50. Autobuffering is
not available on SPORT1 of the ADSP-21msp50 and other mixed-signal
processors with an analog interface.

8 Analog Interface

Before autobuffering is enabled, separate circular buffers must be set up in data memory for the ADC receive and DAC transmit data. This is accomplished by selecting I (index) and M (modify) registers in the analog autobuffer/powerdown register; see Figure 8.5.

Transmit data autobuffered to the DAC is addressed with the I register specified in the ATIREG field (bits 9, 10, 11). Receive data autobuffered from the ADC is addressed with the I register specified in the ARIREG field (bits 4, 5, 6). The modify (M) registers are specified in the ARMREG (bits 2, 3) field and ATMREG (bits 7, 8) field. Since the transfer of ADC and DAC data occurs simultaneously, it is possible to use the same I register for transmit and receive autobuffering. In this case, the buffer is shared for both functions and care should be taken when specifiying a value for the M register.

An autobuffering example program is shown in Listing 8.2.

```
{ ADSP-21msp50 Analog Interface Autobuffer Example    }
{    - configures analog interface                    }
{    - enables analog autobuffer                       }
{    - receive analog data into a 256 word buffer     }
{    - transmit analog data from a 256 word buffer    }

.MODULE/RAM/ABS=0/BOOT=0  auto_example;
.VAR/DM/CIRC  buff1[256];                        {first data buffer}
.VAR/DM/CIRC  buff2[256];                        {second data buffer}
.VAR/DM  flag_bit;                               {tracks buffers}
#define  codec_tx_data  0x3FEC
#define  codec_rx_data  0x3FED
#define  codec_ctrl_reg 0x3FEE
#define  codec_auto_ctrl 0x3FEF

resetv:        JUMP setup; NOP; NOP; NOP;
irq2v:         RTI; NOP; NOP; NOP;               {interrupt vectors ...}
hipwv:         RTI; NOP; NOP; NOP;
hiprv:         RTI; NOP; NOP; NOP;
spt0tv:        RTI; NOP; NOP; NOP;
spt0rv:        RTI; NOP; NOP; NOP;
antv:          RTI; NOP; NOP; NOP;
anrv:          JUMP switch; NOP; NOP; NOP;       {call autobuffer switch}
irq1v:         RTI; NOP; NOP; NOP;
irq0v:         RTI; NOP; NOP; NOP;
timerv:        RTI; NOP; NOP; NOP;
pwrdwnv:       RTI; NOP; NOP; NOP;
```

```
setup:          I0 = ^buff1;              {I0 points to first data buffer}
                L0 = %buff1;
                I1 = ^buff2;              {I1 points to second data buffer}
                L1 = %buff2;
                M0 = 0x1;
                SI = 0x0;                 {initialize flag register}
                DM(flag_bit) = SI;
                                          {use I1 and M0 for tranmsit}
                                          {use I0 and M0 for receive}
                AY0 = 0x0203;             {enable rcv and tx autobuffer}
                DM(codec_auto_ctrl) = AY0;
                AX1 = 0x0060;
                DM(codec_ctrl_reg) = AX1; {power up analog interface}
                IMASK = 0x8;              {enable analog rx interrupt}

wait:           IDLE;                     {wait for autobuffer interrupt}
                JUMP wait;
switch:         AX0 = DM(flag_bit);
                AR = pass AX0;            {check buffer status}
                IF NE JUMP fill_buff2;
fill_buff1:     SI = 0x1;                 {fill buff2 next time}
                AY0 = 0x0013;
                JUMP done;
fill_buff2:     SI = 0x0;                 {fill buff1 next time}
                AY0 = 0x0203;
                JUMP done;
done:           DM(codec_auto_ctrl) = AY0;
                DM(flag_bit) = SI;
                RTI;
.ENDMOD;
```

Listing 8.2 ADSP-21msp50 Analog Autobuffer Program

Receive and transmit autobuffering may be independently enabled
and the two interrupts can occur (and be serviced) independently. This
allows the use of different data buffer lengths when autobuffering both
receive and transmit data. It also allows autobuffering to be used on
only one side, receive or transmit, while the other is serviced at the 8
kHz interrupt rate.

8 Analog Interface

8.5 CIRCUIT DESIGN CONSIDERATIONS

The following sections discuss interfacing analog signals to the ADSP-21msp50.

8.5.1 Analog Signal Input

The analog input signal to the ADSP-21msp50 must be ac-coupled. Figure 8.8 shows the recommended input circuit for the ADSP-21msp50's analog input pin (either VIN$_{NORM}$ or VIN$_{AUX}$). The circuit of Figure 8.8 implements a first-order low pass filter with a 3 dB point at 20 kHz; this is the only filter that must be implemented external to the processor to prevent aliasing of the sampled signal. Since the ADSP-21msp50's sigma-delta ADC uses a highly oversampled approach that transfers the bulk of the anti-aliasing filtering into the digital domain, the off-chip anti-aliasing filter need only be of low order.

In the circuit shown in Figure 8.8, scaling of the analog input is achieved by the resistors R$_{IN}$ and R$_{FB}$. The input signal gain, –R$_{FB}$/R$_{IN}$, can be adjusted from –12 dB to +26 dB by varying the values of these resistors. The ADSP-21msp50's on-chip 20 dB pre-amplifier (of the analog interface) can be enabled when there is not enough gain in the input circuit; the pre-amplifier is configured by bit

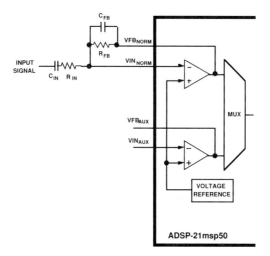

Figure 8.8 Recommended Analog Input Circuit

168

Analog Interface   8

0 (IPS) of the processor's analog control register. Total gain must be configured to ensure that a full-scale input signal (at C_{IN} in Figure 8.8) produces a signal level at the input to the sigma-delta modulator of the ADC that does not exceed V_{INMAX}, which is specified in the *ADSP-21msp50 Data Sheet* under "Analog Interface Electrical Characteristics."

The dc offsetting of the analog input signal is accomplished with the processor's on-chip voltage reference which nominally equals 2.5 V. The input signal must be ac-coupled with an external coupling capacitor (C_{IN}). C_{IN} and R_{IN} should be chosen to ensure a coupling corner frequency of 30 Hz. C_{IN} should be 0.1 µF or larger.

To select values for the components shown in Figure 8.8, use the following equations:

$$\text{Gain} = \frac{-R_{FB}}{R_{IN}}$$

$$C_{IN} = \frac{1}{60\pi R_{IN}}$$

$$C_{FB} = \frac{1}{(2\pi)(20 * 10^3)R_{FB}}$$

$$10 \text{ k}\Omega \leq R_{FB}, R_{IN} \leq 50 \text{ k}\Omega$$

$$150 \text{ pF} \leq C_{FB} \leq 600 \text{ pF}$$

Figure 8.9, which can be found on the following page, shows an example of a typical input circuit configured for 0 dB gain. The circuit's diodes are used to prevent the input signal from exceeding maximum limits.

8 Analog Interface

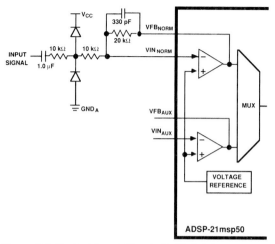

Figure 8.9 Typical Input Circuit (0 dB Gain)

8.5.2 Analog Signal Output

The ADSP-21msp50's differential analog output (VOUTP,VOUTN) is produced by an on-chip differential amplifier which is part of the processor's analog interface. The differential amplifier can drive a minimum load of 2 kΩ ($R_L \geq 2$ kΩ) and has a maximum differential output voltage swing of ±3.156 V peak-to-peak (3.17 dBm0). The differential output can can be ac-coupled directly to a load or dc-coupled to an external amplifier.

Figure 8.10 shows a simple circuit providing a differential output with ac coupling. The capacitor of this circuit (C_{OUT}) is optional; if used, its value can be chosen as follows:

$$C_{OUT} = \frac{1}{(60\pi)R_L}$$

170

Analog Interface   8

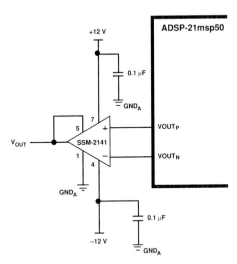

Figure 8.10 Example Circuit For Differential Output With AC Coupling

The VOUTP – VOUTN outputs must be used as differential outputs; do not use either as a single-ended output. Figure 8.11 shows an example circuit which can be used to convert the differential output to a single-ended output. The circuit uses a differential-to-single-ended amplifier, the Analog Devices SSM-2141.

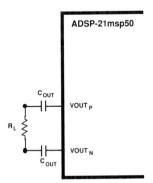

Figure 8.11 Example Circuit For Single-Ended Output

System Interface ◼ 9

9.1   OVERVIEW

This chapter describes the basic system interface features of the ADSP-2100 family processors. The system interface includes various hardware and software features used to control the DSP processor.

Processor control pins include a $\overline{RESET}$ signal, clock signals, flag inputs and outputs, and interrupt requests. The TRAP instruction and $\overline{HALT}$ signal of the ADSP-2100 are described separately.

This chapter describes only the logical relationships of control signals; consult individual processor data sheets for actual timing specifications.

9.2   CLOCK SIGNALS

The ADSP-2101, ADSP-2105, ADSP-2111, and ADSP-21msp50 processors may be operated with a TTL-compatible clock signal input to the CLKIN pin or with a crystal connected between the CLKIN and XTAL pins. In either configuration, CLKIN runs at the instruction cycle rate. If an external clock is used, XTAL must be left unconnected. The CLKIN signal may not be halted or changed in frequency during operation.

Because these processors include an on-chip oscillator circuit, an external crystal can be used. The crystal should be connected between the CLKIN and XTAL pins, with two capacitors connected as shown in Figure 9.1, which can be found on the following page. A parallel-resonant, fundamental frequency, microprocessor-grade crystal should be used. The frequency value selected for the crystal should be equal to the desired instruction rate for the processor. For example, a 16 MHz crystal is used for a processor with an instruction rate of 16 MHz. The internal phased lock loop of the processor generates an internal clock which is four times the instruction rate.

Unlike the other processors of the family, the ADSP-2100 processor does not have an on-chip oscillator circuit and therefore has no XTAL pin. The ADSP-2100 must be operated with a TTL-compatible clock signal input to CLKIN. CLKIN runs at four times the instruction cycle rate: for a 10 MHz instruction cycle rate, the processor requires a 40 MHz clocking signal.

173

9 System Interface

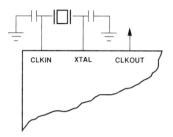

Figure 9.1 External Crystal Connections

The processors also generate a CLKOUT signal which is synchronized to the processors' internal cycles and operates at the instruction cycle rate. In all processors except the ADSP-2100, a phase-locked loop is used to generate CLKOUT and to divide each instruction cycle into a sequence of internal time periods called processor states. The relationship between the phases of CLKIN, CLKOUT and the processor states is shown in Figure 9.2 (for the ADSP-2101, ADSP-2105, ADSP-2111, and ADSP-21msp50 processors). The phases of the internal processor clock are dependent upon the period of the external clock.

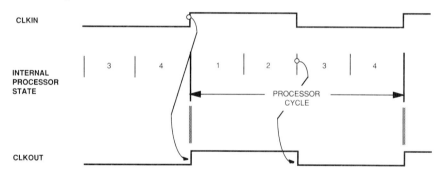

Figure 9.2 Clock Signals & Processor States

The ADSP-2100 processor uses the phases of CLKIN to generate CLKOUT and its internal states, as shown in Figure 9.3. Note that the ADSP-2100 distinguishes eight internal states per cycle, whereas the other family processors distinguish only four. In the ADSP-2100, the eighth state of each cycle provides a neutral halting point for the processor when operation is suspended. In the other processors, the rising edge of CLKOUT always marks the transition between state four of one cycle and state one of the next cycle.

174

System Interface   9

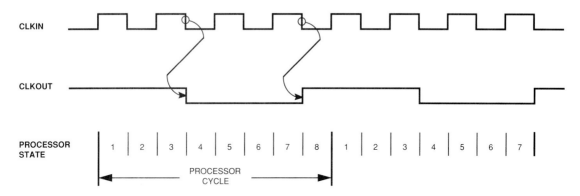

Figure 9.3 Clock Signals & Processor States (ADSP-2100)

9.2.1   Synchronization Delay

Each processor has several asynchronous inputs (interrupt requests, for example), which can be asserted in arbitrary phase to the processor clock. The processor synchronizes such signals before recognizing them. The delay associated with signal recognition is called synchronization delay.

Different asynchronous inputs are recognized at different points in the processor cycle. Any asynchronous input must be valid prior to the recognition point to be recognized in a particular cycle. If an input does not meet the setup time on a given cycle, it is recognized either in the current cycle or during the next cycle if it remains valid.

In family processors other than the ADSP-2100, edge-sensitive interrupt requests are latched internally so that the request signal only has to meet the pulse width requirement. To ensure the recognition of any asynchronous input, however, the input must be asserted for at least one full processor cycle plus setup and hold time. Setup and hold times are specified in the data sheet for each individual device.

9.2.2   1x Frequency Clock Considerations

Each processor (except the ADSP-2100) requires only a 1X frequency clock signal. They use what is effectively an on-chip phase-locked loop to generate the higher frequency internal clock signals and CLKOUT. Because these clocks are generated based on the rising edge of CLKIN, there is no ambiguity about the phase relationship of two processors sharing the same input clock. Multiple processor synchronization is simplified as a result.

175

9   System Interface

Using a 1X frequency input clock with the phase-locked loop to generate the various internal clocks imposes certain restrictions. The CLKIN signal must be valid long enough to achieve phase lock before RESET can be de-asserted. Also, the clock frequency cannot be changed unless the processor is in RESET. Refer to the processor data sheets for details.

9.3   RESET

RESET halts execution and causes a hardware reset of the ADSP-2100 family processors. The RESET signal must be asserted when the processor is powered up to assure proper initialization.

9.3.1   Reset On All Family Processors (Except ADSP-2100)

For the ADSP-2101, ADSP-2105, ADSP-2111, and ADSP-21msp50 processors, Tables 9.1–9.4 show the RESET state of various registers, including the processors' on-chip memory-mapped status/control registers. The values of any registers not listed are undefined at reset. The contents of on-chip memory (DM and PM RAM) are unchanged after RESET, except as shown in Tables 9.1–9.4 for the data memory-mapped control/status registers. The CLKOUT signal continues to be generated by the processor during RESET.

In all family processors, the contents of the computation unit (ALU, MAC, Shifter) and data address generator (DAG1, DAG2) registers are undefined following RESET.

When RESET is released, the boot memory booting operation takes place, depending on the state of the processor's MMAP pin. (Program booting is described in Chapter 10, "Memory Interface.") Certain processors that utilize internal program memory ROM will not boot.

For the ADSP-2111 and ADSP-21msp50 processors, which include a host interface port, setting the software reset bit in the HSR7 register has the same affect as asserting RESET. This allows either the host processor or the ADSP-2111/ADSP-21msp50 itself to initiate a reset under software control.

In a multiprocessing system using several processors, a synchronous RESET is required.

System Interface   9

9.3.2   Reset On ADSP-2100 Processor

In the ADSP-2100, $\overline{RESET}$ performs the following:

- Initializes internal clock circuitry
- Resets all internal stack pointers
- Clears the cache memory monitor and invalidates cache contents
- Clears IMASK register, masking all interrupts (ICNTL is undefined)
- Clears MSTAT register

The contents of all ADSP-2100 computation unit (ALU, MAC, Shifter) and data address generator (DAG1, DAG2) registers are undefined following $\overline{RESET}$. A $\overline{RESET}$ on the ADSP-2100 is recognized on any rising edge of CLKIN, and must be asserted for at least four CLKIN cycles. The ADSP-2100 remains in state 4 during $\overline{RESET}$ (see Figure 9.3). When $\overline{RESET}$ is released, the ADSP-2100 goes from state 4 to state 5 on the second rising edge of CLKIN following release. Refer the *ADSP-2100 Data Sheet* for exact timing information. The CLKOUT signal is stopped during $\overline{RESET}$.

9.4   SOFTWARE-FORCED REBOOTING

Software-forced reboots from boot memory can be accomplished in all family processors except the ADSP-2100. Setting the BFORCE bit in the processor's system control register causes a reboot. A software forced reboot clears the context of the processor and initializes some registers.

Tables 9.1–9.4 show the state of the processor registers after a software-forced reboot. The values of any registers not listed are unchanged by a reboot.

During booting (and rebooting), all interrupts including serial port interrupts are masked and autobuffering is disabled. The serial port(s) remain active; one transfer—from internal shift register to data register—can occur for each serial port before there are overrun problems.

The timer runs during a reboot. If a timer interrupt occurs during the reboot, it is masked. Thus, if more than one timer interrupt occurs during the reboot, the processor latches only the first. A timer overrun can occur.

9 System Interface

Control Field	Description	Reset	Reboot
Bus Exchange Register			
PX	PX register	undefined	undefined
Status Registers			
IMASK	Interrupt service enables	0	0
ASTAT	Arithmetic status	0	0
MSTAT	Mode status	0	unchanged
SSTAT	Stack status	0x55	0x55
ICNTL	Interrupt control	undefined	unchanged
IFC	Interrupt force/clear	0	0
Control Registers (memory-mapped)			
BWAIT	Boot memory wait states	3	unchanged
BPAGE	Boot page	0	unchanged
SPORT1 configure	Configuration	1	unchanged
SPE0	SPORT0 enable	0	unchanged
SPE1	SPORT1 enable	0	unchanged
DWAIT0–4	Data memory wait states	7	unchanged
PWAIT	Program memory wait	7	unchanged
TCOUNT	Timer count register	undefined	operates during reboot
TPERIOD	Timer period register	undefined	unchanged
TSCALE	Timer scale register	undefined	unchanged
Serial Port Control Registers (memory-mapped, one set per SPORT)			
ISCLK	Internal serial clock	0	unchanged
RFSR, TFSR	Frame sync required	0	unchanged
RFSW, TFSW	Frame sync width	0	unchanged
IRFS, ITFS	Internal frame sync	0	unchanged
INVRFS, INVTFS	Invert frame sense	0	unchanged
DTYPE	Companding type, format	0	unchanged
SLEN	Serial word length	0	unchanged
SCLKDIV	Serial clock divide	undefined	unchanged
RFSDIV	RFS divide	undefined	unchanged
Multichannel word enable bits		undefined	unchanged
MCE	Multichannel enable	0	unchanged
MCL	Multichannel length	0	unchanged
MFD	Multichannel frame delay	0	unchanged
INVTDV	Invert transmit data valid	0	unchanged
RBUF, TBUF	Autobuffering enable	0	0
TIREG, RIREG	Autobuffer I index	undefined	unchanged
TMREG, RMREG	Autobuffer M index	undefined	unchanged
FO *(SPORT1 only)*	Flag Out value	undefined	unchanged

Table 9.1 ADSP-2101 State After Reset Or Software Reboot

System Interface   9

Control Field	Description	Reset	Reboot
Bus Exchange Register			
PX	PX register	undefined	undefined
Status Registers			
IMASK	Interrupt service enables	0	0
ASTAT	Arithmetic status	0	0
MSTAT	Mode status	0	unchanged
SSTAT	Stack status	0x55	0x55
ICNTL	Interrupt control	undefined	unchanged
IFC	Interrupt force/clear	0	0
Control Registers (memory-mapped)			
BWAIT	Boot memory wait states	3	unchanged
BPAGE	Boot page	0	unchanged
SPORT1 configure	Configuration	1	unchanged
SPE1	SPORT1 enable	0	unchanged
DWAIT0–4	Data memory wait states	7	unchanged
PWAIT	Program memory wait	7	unchanged
TCOUNT	Timer count register	undefined	operates during reboot
TPERIOD	Timer period register	undefined	unchanged
TSCALE	Timer scale register	undefined	unchanged
Serial Port 1 Control Registers (memory-mapped)			
ISCLK	Internal serial clock	0	unchanged
RFSR, TFSR	Frame sync required	0	unchanged
RFSW, TFSW	Frame sync width	0	unchanged
IRFS, ITFS	Internal frame sync	0	unchanged
INVRFS, INVTFS	Invert frame sense	0	unchanged
DTYPE	Companding type, format	0	unchanged
SLEN	Serial word length	0	unchanged
SCLKDIV	Serial clock divide	undefined	unchanged
RFSDIV	RFS divide	undefined	unchanged
RBUF, TBUF	Autobuffering enable	0	0
TIREG, RIREG	Autobuffer I index	undefined	unchanged
TMREG, RMREG	Autobuffer M index	undefined	unchanged
FO	Flag Out value	undefined	unchanged

Table 9.2 ADSP-2105 State After Reset Or Software Reboot

179

9   System Interface

Control Field	Description	Reset	Reboot
Bus Exchange Register			
PX	PX register	undefined	undefined
Status Registers			
IMASK	Interrupt service enables	0	0
ASTAT	Arithmetic status	0	0
MSTAT	Mode status	0	unchanged
SSTAT	Stack status	0x55	0x55
ICNTL	Interrupt control	undefined	unchanged
IFC	Interrupt force/clear	0	0
Control Registers (memory-mapped)			
BWAIT	Boot memory wait states	3	unchanged
BPAGE	Boot page	0	unchanged
SPORT1 configure	Configuration	1	unchanged
SPE0	SPORT0 enable	0	unchanged
SPE1	SPORT1 enable	0	unchanged
DWAIT0–4	Data memory wait states	7	unchanged
PWAIT	Program memory wait	7	unchanged
TCOUNT	Timer count register	undefined	operates during reboot
TPERIOD	Timer period register	undefined	unchanged
TSCALE	Timer scale register	undefined	unchanged
Serial Port Control Registers (memory-mapped, one set per SPORT)			
ISCLK	Internal serial clock	0	unchanged
RFSR, TFSR	Frame sync required	0	unchanged
RFSW, TFSW	Frame sync width	0	unchanged
IRFS, ITFS	Internal frame sync	0	unchanged
INVRFS, INVTFS	Invert frame sense	0	unchanged
DTYPE	Companding type, format	0	unchanged
SLEN	Serial word length	0	unchanged
SCLKDIV	Serial clock divide	undefined	unchanged
RFSDIV	RFS divide	undefined	unchanged
Multichannel word enable bits		undefined	unchanged
MCE	Multichannel enable	0	unchanged
MCL	Multichannel length	0	unchanged
MFD	Multichannel frame delay	0	unchanged
INVTDV	Invert transmit data valid	0	unchanged
RBUF, TBUF	Autobuffering enable	0	0
TIREG, RIREG	Autobuffer I index	undefined	unchanged
TMREG, RMREG	Autobuffer M index	undefined	unchanged
FO *(SPORT1 only)*	Flag Out value	undefined	unchanged

System Interface   9

Host Interface Port Registers (memory-mapped)

HDR0-5	HIP data registers	undefined	used during HIP reboot
HSR6	HIP status register	0x0000	used during HIP reboot
HSR7	HIP status register	0x0080	unchanged
HMASK	HIP interrupt enables	0	unchanged

Table 9.3 ADSP-2111 State After Reset Or Software Reboot

Control Field	Description	Reset	Reboot
Bus Exchange Register			
PX	PX register	undefined	undefined
Status Registers			
IMASK	Interrupt service enables	0	0
ASTAT	Arithmetic status	0	0
MSTAT	Mode status	0	unchanged
SSTAT	Stack status	0x55	0x55
ICNTL	Interrupt control	undefined	unchanged
IFC	Interrupt force/clear	0	0
Control Registers (memory-mapped)			
BWAIT	Boot memory wait states	3	unchanged
BPAGE	Boot page	0	unchanged
SPORT1 configure	Configuration	1	unchanged
SPE0	SPORT0 enable	0	unchanged
SPE1	SPORT1 enable	0	unchanged
DWAIT0–4	Data memory wait states	7	unchanged
PWAIT	Program memory wait	7	unchanged
TCOUNT	Timer count register	undefined	operates during reboot
TPERIOD	Timer period register	undefined	unchanged
TSCALE	Timer scale register	undefined	unchanged
ROMENABLE	Program memory ROM enable bit	0	unchanged
XTALDIS	Powerdown XTAL drive enable	0	unchanged
XTALDELAY	Powerdown XTAL drive delay	0	unchanged
Serial Port Control Registers (memory-mapped, one set per SPORT)			
ISCLK	Internal serial clock	0	unchanged
RFSR, TFSR	Frame sync required	0	unchanged
RFSW, TFSW	Frame sync width	0	unchanged
IRFS, ITFS	Internal frame sync	0	unchanged
INVRFS, INVTFS	Invert frame sense	0	unchanged
DTYPE	Companding type, format	0	unchanged
SLEN	Serial word length	0	unchanged
SCLKDIV	Serial clock divide	undefined	unchanged

9 System Interface

RFSDIV	RFS divide	undefined	unchanged
Multichannel word enable bits		undefined	unchanged
MCE	Multichannel enable	0	unchanged
MCL	Multichannel length	0	unchanged
MFD	Multichannel frame delay	0	unchanged
INVTDV	Invert transmit data valid	0	unchanged
RBUF, TBUF	Autobuffering enable	0	0
TIREG, RIREG	Autobuffer I index	undefined	unchanged
TMREG, RMREG	Autobuffer M index	undefined	unchanged
FO *(SPORT1 only)*	Flag Out value	undefined	unchanged
Host Interface Port Registers (memory-mapped)			
HDR0-5	HIP data registers	undefined	used during HIP reboot
HSR6	HIP status register	0x0000	used during HIP reboot
HSR7	HIP status register	0x0080	unchanged
HMASK	HIP interrupt enables	0	unchanged
Analog Autobuffer/Powerdown Registers			
ARBUF	Receive autobuffer enable	0	0
ATBUF	Transmit autobuffer enable	0	0
control bits	Analog autobuffer control bits	0	unchanged

Table 9.4 ADSP-21msp50 State After Reset Or Software Reboot

9.5   EXTERNAL INTERRUPTS

Each ADSP-2100 family processor supports one or more prioritized, individually maskable external interrupts that can be either level- or edge-triggered. The interrupt request pins are named $\overline{IRQ0}$, $\overline{IRQ1}$, etc.

The ADSP-2100 processor has four interrupt request pins: $\overline{IRQ0}$, $\overline{IRQ1}$, $\overline{IRQ2}$, and $\overline{IRQ3}$. The ADSP-2101, ADSP-2105, ADSP-2111, and ADSP-21msp50 processors all include an $\overline{IRQ2}$ pin; on these processors, the (optional) alternate configuration of SPORT1 also provides $\overline{IRQ0}$ and $\overline{IRQ1}$ pins. The configuration of SPORT1 as either a serial port or as interrupts (and flags) is determined by bit 10 of the processor's system control register.

Internal interrupts of the ADSP-2101, ADSP-2105, ADSP-2111, and ADSP-21msp50 processors, including serial port, timer, host interface port, and analog interface interrupts, are discussed in other chapters.

Additional information about interrupt masking, set up, and operation can be found in Chapter 3, "Program Control."

System Interface 9

9.5.1 Interrupt Sensitivity

Individual external interrupts can be configured in the ICNTL register as either level-sensitive or edge-sensitive.

Level-sensitive interrupts operate by asserting the interrupt request line ($\overline{IRQx}$) until the request is recognized by the processor. Once recognized, the request must be de-asserted before unmasking the interrupt so that the interrupt is not serviced twice.

In contrast, edge-triggered interrupt requests are latched when any high-to-low transition occurs on the interrupt line. The processor latches the interrupt so that the request line may be held at any level for an arbitrarily long period between interrupts. This latch is automatically cleared when the interrupt is serviced. Edge-triggered interrupts require less external hardware than level-sensitive requests since there is never a need to hold or negate the request. With level-sensitive interrupts, however, many interrupting devices can share a single request input; this allows easy system expansion.

An interrupt request will be serviced if it is not masked (in the IMASK register) and a higher priority request is not pending. Valid requests initiate an interrupt servicing sequence that vectors the processor to the appropriate interrupt vector address. The interrupt vector addresses for each family processor are given in Table 3.2 in Chapter 3. There is a synchronization delay associated with both external interrupt request lines and internal interrupts.

On processors other than the ADSP-2100, if an interrupt occurs during the extra cycles required to execute an instruction that accesses external memory more than once, it is not recognized between the cycles, only before or after. Interrupts are latched, but not serviced, during bus grant ($\overline{BG}$) unless the GO mode is enabled. GO mode is not available on the ADSP-2100 processor; interrupts are latched, but not serviced, during $\overline{HALT}$, TRAP, and bus grant ($\overline{BG}$), and also while the processor is waiting for DMACK.

In order to service an interrupt, the processor must be running and executing instructions. On processors other than the ADSP-2100, the IDLE instruction can be used to effectively halt processor operations while waiting for an interrupt.

Edge-sensitive and level-sensitive interrupt requests are serviced similarly. Edge-sensitive interrupts may remain active (low) indefinitely, while level-sensitive interrupts must be de-asserted before the RTI instruction is executed; otherwise, the same interrupt immediately recurs.

183

9 System Interface

Care must be taken with processors that have a serial port which can be configured for alternate function (i.e. $\overline{IRQ0}$ and $\overline{IRQ1}$). If the RFS1 or TFS1 input is held low when configured for the serial port function and the serial port is reconfigured for alternate function, an interrupt request can be generated. This interrupt request can be cleared with the use of the IFC register.

9.6 FLAG PINS

All family processors except the ADSP-2100 provide flag pins. The alternate configuration of SPORT1 includes a Flag In (FI) pin and a Flag Out (FO) pin. The configuration of SPORT1 as either a serial port or as flags and interrupts is determined by bit 10 of the processor's system control register.

FI can be used to control program branching, using the IF FLAG_IN and IF NOT FLAG_IN conditions of the JUMP and CALL instructions. These conditions are evaluated based on the last state of the FI pin; FLAG_IN is true if FI was last sampled as a 1 and false if last sampled as a 0. FO can be used as a general purpose external signal. The state of FO is also available as a read-only bit of the SPORT1 control register.

The ADSP-2111 and ADSP-21msp50 processors each have three additional flag output pins: FL0, FL1 and FL2. These flags (and FO) can be controlled in software to signal events or conditions to any external device such as a host processor. The Modify Flag Out instruction, which is conditional, can perform SET, RESET and TOGGLE actions—this instruction allow programs executing on the DSP processor to control the state of its flag output pins. Note that if the condition in the Modify Flag Out instruction is CE (counter expired), the counter is not decremented as in other IF CE instructions.

Flag outputs FL0, FL1 and FL2 are set to 1 at $\overline{RESET}$. The Flag Out (F0) is not affected by $\overline{RESET}$.

9.7 HALT & TRAP (ADSP-2100 ONLY)

The ADSP-2100 processor includes a $\overline{HALT}$ input which is used to temporarily suspend processor operation.

The $\overline{HALT}$ input is recognized at the end of state three of the processor cycle. The ADSP-2100 will halt in state eight of the current cycle if it was performing a program memory instruction fetch when the $\overline{HALT}$ was recognized, or in state eight of the following cycle if it was performing a

System Interface   9

program memory data fetch. In the latter case, the second cycle is a forced program memory instruction fetch even if the instruction is available from cache. Hence, the processor is always halted on a program memory instruction fetch and the controlling device can observe the address (on the PMA bus) where execution was terminated.

Normal processor operation is resumed when the $\overline{\text{HALT}}$ line is released. You must ensure that DMACK is high when $\overline{\text{HALT}}$ is released.

If $\overline{\text{HALT}}$ is asserted during a bus grant, it is latched but not serviced until the ADSP-2100 regains control of its busses; the processor is already halted by the bus grant. Likewise, if $\overline{\text{HALT}}$ is asserted when the processor is waiting for DMACK, it is latched but not serviced. When the processor is halted due to the $\overline{\text{HALT}}$ input, $\overline{\text{BR}}$ is latched and serviced normally.

The TRAP signal is generated by the ADSP-2100 whenever a TRAP instruction is executed. It is asserted on the transition between states seven and eight of that cycle and the processor is halted in state eight. The PMA bus provides the address of the instruction that follows the TRAP instruction. The TRAP signal will remain active until $\overline{\text{HALT}}$ is asserted. Upon recognition of $\overline{\text{HALT}}$, the ADSP-2100 releases TRAP but remains in a halt state.

If $\overline{\text{BR}}$ is asserted during TRAP it is latched and serviced normally.

Normal operation is resumed when $\overline{\text{HALT}}$ is released, as shown in Figure 9.4.

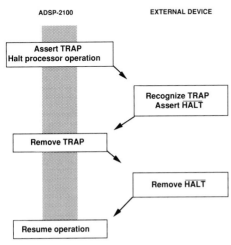

ADSP-2100   EXTERNAL DEVICE

Assert TRAP
Halt processor operation

Recognize TRAP
Assert $\overline{\text{HALT}}$

Remove TRAP

Remove $\overline{\text{HALT}}$

Resume operation

Figure 9.4 TRAP Sequence

9 System Interface

9.8 POWERDOWN (ADSP-21MSP5X ONLY)

The ADSP-21msp5x processors include a low power feature that allows the processor to enter a very low power dormant state through hardware or software control. In this CMOS standby state, power consumption is less than 1 mW. This feature is useful for applications, such as battery powered operation, where power conservation is necessary. Features of powerdown include:

- Internal clocks are disabled

- Processor registers and memory contents are maintained

- Ability to recover from powerdown in less than 100 clock cycles

- Ability to disable internal oscillator when using crystal

- No need to shut down clock for lowest power when using external oscillator

- Interrupt support for executing "housekeeping" code before entering powerdown and after recovering from powerdown

- User selectable powerup context

Even though the processor is put into the powerdown mode, the lowest level of power consumption still might not be achieved if the proper guidelines are not followed. Lowest possible power consumption requires no additional current flow through processor output pins and no switching activity on active input pins. Therefore, a careful analysis of pin loading in your circuit is required. The following sections detail the proper powerdown procedure as well as provide guidelines for clock and output pin connections required for optimum low-power performance.

9.8.1 Powerdown Control

You can control several parameters of powerdown operation through control bits of the Analog Autobuffer/Powerdown Control Register. This control register is memory-mapped at location 0x3FEF and is shown in Figure 9.5.

System Interface 9

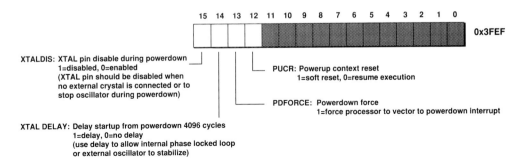

Figure 9.5 Analog Autobuffer/Powerdown Control Register

9.8.2   Entering Powerdown

The powerdown sequence is defined as follows.

- Initiate the powerdown sequence by applying a high to low transition to the PWD pin or by setting the powerdown force control bit in the Analog Autobuffer/Powerdown Control Register (memory-mapped location 0x3FEF, bit 13.)

- The processor vectors to the non-maskable powerdown interrupt vector located at 0x002C. **Note:** The powerdown interrupt is never masked. You must be careful not to cause multiple powerdown interrupts to occur or stack overflow could result. Multiple powerdown interrupts can occur if the PWD input is pulsed while the processor is servicing the powerdown interrupt or if there is significant noise present.

- Any number of housekeeping instructions, starting at location 0x002C, can be executed prior to the processor entering the powerdown mode. Typically, this section of code is used to configure the powerdown state, disable on-chip peripherals such as the analog interface and clear pending interrupts.

- The processor enters the powerdown mode when the processor executes an IDLE instruction. The processor may take either one or two cycles to powerdown depending upon internal clock states during the execution of the IDLE instruction. All register and memory contents are maintained while in powerdown. Also, all active outputs are held in whatever state they are in before going into powerdown.

9   System Interface

If an RTI is executed before the IDLE instruction, then the processor returns from the powerdown interrupt and the powerdown sequence is aborted.

While the processor is in the powerdown mode, the processor is in CMOS standby. This allows the lowest level of power consumption where most input pins are ignored. Active inputs need to be held at CMOS levels to achieve lowest power. More information can be found in section 9.8.5, "Operation During Powerdown."

9.8.3   Exiting Powerdown

The powerdown mode can be exited with the use of the $\overline{PWD}$ pin or with RESET. There are also several user-selectable modes for start-up from powerdown which specify a start-up delay as well as specify the program flow after start-up. This allows the program to resume from where it left off before powerdown or for the program context to be cleared.

9.8.3.1   Ending Powerdown With The $\overline{PWD}$ Pin

Applying a low-to-high transition to the $\overline{PWD}$ pin will take the processor out of powerdown mode. You have the option of selecting the amount of time the processor takes to come out of the powerdown mode with the "delay start-up from powerdown" control bit (XTALDELAY, bit 14 in the Powerdown Control Register.) If this bit is cleared to 0, no additional delay over the quick start-up (100 cycles) is introduced. If this bit is set to 1, a delay of 4096 cycles is introduced. The delay feature is used depending upon the state of an external clock oscillator at the time of powerup or if the internal clock is disabled. This is further discussed in the sections "Systems Using an External TTL/CMOS Clock and Systems Using The Internal Oscillator."

You can also program one of two options directing the processor how to resume operation. The context for exiting powerdown is set by bit 12 (exit from powerup context reset bit) of the Powerdown Control Register.

If the control bit named "powerup context reset" is cleared to 0, the processor will continue to execute instructions following the IDLE instruction. For example, a high-to-low transition is applied to the $\overline{PWD}$ pin which causes the processor to vector to the powerdown interrupt routine. In this routine, a few housekeeping tasks are performed and the IDLE instruction is executed. The processor powers down. Some time later a low to high transition is applied to the $\overline{PWD}$ pin causing the processor to exit powerdown mode. Since the "powerup context reset" bit is set to a 0, the processor resumes executing instructions in the powerdown interrupt routine with the instruction after the IDLE instruction which caused the powerdown. When an RTI instruction is encountered, control then passes back to the main routine.

If the powerup context reset bit is set to 1, the processor resumes operation from powerdown by clearing the PC, STATUS, LOOP and CNTR stack. The IMASK and ASTAT registers are set to 0 and the SSTAT goes to 0x55. The processor will start executing instructions from address zero.

9.8.3.2   Ending Powerdown With The $\overline{RESET}$ Pin

If $\overline{RESET}$ is asserted while the processor is in the powerdown mode, the processor is reset and instructions are executed from address 0x0000. A boot is performed if the MMAP pin is set to 0. If the $\overline{RESET}$ pin is used to exit powerdown, then it must be held low for the appropriate number of cycles. If the clock is stopped at powerup or operating at a different frequency at powerup than it was before powerdown, $\overline{RESET}$ must be held long enough for the oscillator to stabilize plus an additional 1000 clock cycles for the phase locked loop to lock. The time required for the oscillator to stabilize depends upon the type of crystal used and capacitance of the external crystal circuit. Typically 2000 cycles is adequate for clock stabilization time.

If the clock was not stopped at powerup and is at a stable frequency at powerup (same as before powerdown), only 5 cycles of $\overline{RESET}$ is required.

When ending powerdown with $\overline{RESET}$, the "delay start-up from powerdown" control bit is ignored.

9.8.4   Start-up Time After Powerdown

The time required to exit the powerdown state depends on whether an internal or external oscillator is used, and the method used to exit powerdown.

9.8.4.1   Systems Using An External TTL/CMOS Clock

When the processor is in powerdown, the external clock signal is ignored. Therefore it is not necessary to stop the external clock since no power is wasted while the external clock is running.

Bit 15, XTALDIS, of the powerdown control register (XTAL pin drive during powerdown) should always be set before entering powerdown. This specifies that the XTAL pin is not to be driven by the processor. There is no need to drive the XTAL pin when an external oscillator is used. Disabling the XTAL drive saves power.

9 System Interface

After the processor is taken out of the powerdown mode by either the PWD pin or RESET, it will begin executing instructions after a maximum start-up time of 100 cycles as long as the clock oscillator is stable and at the same frequency as before powerdown.

If the external clock is unstable when the processor exits powerdown, then the "delay start-up from powerdown" control bit can be set. This allows time for the external clock to stabilize by inserting an additional 4096 cycle delay before the processor starts to execute instructions. The start-up delay can only be used when the processor is taken out of powerdown mode with the PWD pin.

If the processor is taken out of powerdown by RESET and the clock is stable and at the same frequency as before powerdown, RESET needs to be held for only 5 cycles.

9.8.4.2 Systems Using A Crystal And The Internal Oscillator

A trade-off can be made so that a fast start-up is possible yet power is being consumed by leaving the oscillator running during powerdown. If a fast start-up is desired, then you must clear bits 14 and 15 of the powerdown control register to 0 before entering powerdown. This selects no additional delay after start-up from powerdown and drives the external crystal during powerdown. In this configuration, the oscillator will continue to operate and the processor will start executing instructions in less than 100 cycles after the low to high signal transition at the PWD pin. The XTAL pin will also be driven and the powerdown power consumption will be higher than the 1 mW specification. The following code example shows the powerdown interrupt routine.

```
{ Sample Powerdown Code                                       }
{ Located at Address 0x002C                                    }
    pwd_int: ax0 = 0x0000; { enable crystal, no delay }
             dm(0x3FEF) = ax0;
             idle;
             rti;
```

If lowest possible power consumption is required, then you must set bits 14 and 15 of the powerdown control register to 1 before entering powerdown. This selects an additional 4096 cycle delay to allow the oscillator to start and the phase locked loop to lock after start-up and disables the drive to the XTAL pin during powerdown. The following code example shows the powerdown interrupt routine.

```
{ Sample Powerdown Code                                    }
{ Located at Address 0x002C                                }
    pwd_int: ax0 = 0xC000; { disable crystal, delay }
             dm(0x3FEF) = ax0;
             idle;
             rti;
```

Depending upon the particular situation and external system conditions, the powerdown modes shown above could be set conditionally. If you want to powerdown for a long time you may want to set the mode for lowest power consumption. If you want to powerdown for a short time, lowest power consumption may not be that important.

If the RESET pin is used to exit powerdown and the clock has been stopped, then RESET must be held low for 1000 clock cycles plus the time required for the phase locked loop to lock and the crystal oscillator to stabilize (typically 2000 cycles.) If the clock is running during powerup, a RESET signal of only 5 cycles is required.

9.8.5   Operation During Powerdown

Some processor circuitry may still be active during powerdown mode. Also, some output pins remain active. A good understanding of these states will allow you to determine the best low-power configuration for your system. By keeping output loading and input switching to a minimum the lowest possible power consumption can be achieved.

9.8.5.1 Analog Interface

You must powerdown the analog interface separately from the processor, as described in section 8.4.1.1, "Analog Control Register." The analog interface does not work during powerdown and causes additional power to be dissipated if not disabled. The following code example shows a powerdown interrupt routine for the ADSP-21msp50.

```
{ Sample Powerdown Code                                        }
{ Located at Address 0x002C                                    }
    pwd_int: ax0 = 0x0000; { powerdown analog interface }
             dm(0x3FEE) = ax0;
             ax0 = 0x0000; { enable crystal, no delay   }
             dm(0x3FEF) = ax0;
             NOP;
             idle;
             rti;
```

9   System Interface

It takes three cycles for the analog section to powerdown. The IDLE instruction should not be executed before these three cycles have elapsed.

9.8.5.2   SPORTS

The circuitry of the serial ports is not directly affected by powerdown. The SPORTs are indirectly affected if an internally generated SCLK or frame sync is required. SPORT circuitry continues to operate during powerdown.

It is possible to clock data into or out of the serial ports during powerdown. You must supply an external serial clock to support operation during powerdown. No interrupts or autobuffer operations will be serviced during powerdown. Instead, the SPORT interrupts are latched and can be serviced if the processor exits powerdown without resetting the processor. Data clocked into the processor will remain in the receive (RX) registers. Autobuffer transfers will occur after the device exits powerdown if the processor is not powered up with $\overline{\text{RESET}}$. Note that any SPORT activity will increase the power consumption above the 1 mW specification.

If an external serial clock and an external frame sync signal are supplied, data can be clocked into the RX register or out of the TX register during powerdown. Since the TX register can not be updated while the processor is in powerdown, the same value is repeatedly clocked out the serial port. Also, data in the RX register is continually overwritten since the RX register can not be read by the processor during powerdown.

If an external serial clock is used with an internal frame sync, frame sync signals continue to be generated during powerdown since they are derived from the the the serial clock. Data bits continue to be received with the RX register being overwritten. Since data is only transmitted when the TX register is written, data bits are only transferred out of the processor if the processor is put in powerdown during a serial port transfer. The serial port transfer in progress, while the processor is put into powerdown, is allowed to complete. Since an internal frame sync is used, no subsequent transmit frame syncs are generated while in powerdown.

If internal serial clock is used, there is no SPORT activity during powerdown; the serial clock stops.

Lowest power dissipation is achieved when active SPORT pins are not changing during powerdown and are held at CMOS levels.

System Interface   9

9.8.5.3   HIP During Powerdown

The circuitry of the HIP is not directly affected by powerdown. The HIP is indirectly affected since the processor, when in powerdown, is unable to service interrupts or read and write HIP data registers. HIP circuitry continues to operate during powerdown.

The host can write to the HIP register during powerdown but the processor is disabled and cannot service interrupts. Instead, HIP interrupts are latched and can be serviced if the processor exits powerdown without resetting the processor.

If the HDR overwrite bit (bit 7 in HSR7) is cleared, a host acknowledge signal will not be asserted until the processor has read data written by the host. During powerdown, the processor is unable to read the data register and the host acknowledge signal will not be asserted. Care must be taken in a system where the host waits for a host acknowledge. In this case, it is possible that the host will "hang" waiting for the acknowledge while the DSP processor is in powerdown.

While in powerdown, the processor can be reset by writing the HSR software reset bit. This will produce the same results as asserting the RESET pin for five cycles (minimum RESET pulse) on the processor. If an external crystal is used and the clock has been stopped, this reset duration is too short; therefore software reset cannot be used in this mode. Note that any HIP activity will increase the power consumption above the 1 mW specification.

Two mode pins, HMD0 and HMD1, are used to put the processor's HIP into one of four possible modes. When HMD0 = 1, the HIP data bus is multiplexed for both address and data. In this case, the HIP data bus inputs are active during powerdown and any bus activity will result in higher power dissipation. Also, inputs must be at CMOS levels. If this host mode is used and there is potential for the bus to be floating, pull-up resistors should be used on the data lines. If you desire the host to communicate with other devices on the bus while the DSP processor is in powerdown, HMD0 should be held low to avoid extra power to be dissipated. When the HIP is put in other modes where data inputs are not active this is not a problem.

Lowest power dissipation is achieved when the HIP pins are not changing during powerdown and are held at CMOS levels.

9 System Interface

9.8.5.4 *Interrupts And Flags*

Interrupts are latched and can be serviced if the processor exits powerdown without resetting the processor. Any activity on the interrupt or flag in pins during powerdown will increase the power consumption. There should also be no resistive load on the flag out pins (as with any active output pin) if lowest power is desired.

9.8.6 Conditions For Lowest Power Consumption

The state of all processor pins during powerdown is shown in the Table 9.5.

Pin Name	Direction	Status During Powerdown
$\overline{\text{RESET}}$	I	Active
$\overline{\text{PWD}}$	I	Active
$\overline{\text{IRQ2}}$	I	Active, latched but not serviced
MMAP	I	Active
$\overline{\text{BR}}$	I	Active, no response until after powerdown
$\overline{\text{BG}}$	O	Driven HIGH unless bus is granted
CLKIN	I	Input buffer inactive, but XTAL oscillator is active unless XTALDIS bit is set
CLKOUT	O	Driven HIGH
XTAL	O	Driven HIGH if XTALDIS set, inversion of CLKIN otherwise
PWDACK	O	Driven HIGH
$\overline{\text{PMS}}$	O	Driven HIGH, high impedance if bus granted
$\overline{\text{DMS}}$	O	Driven HIGH, high impedance if bus granted
$\overline{\text{BMS}}$	O	Driven HIGH, high impedance if bus granted
$\overline{\text{RD}}$	O	Driven HIGH, high impedance if bus granted
$\overline{\text{WR}}$	O	Driven HIGH, high impedance if bus granted
$\overline{\text{PMS}}$	O	Driven HIGH, high impedance if bus granted
ADDR<13:0>	O	High impedance
DATA<23:0>	I	Inactive
DATA<23:0>	O	High impedance
SCLK0	I	Active
SCLK0	O	Driven to static level if internal, high impedance otherwise
TFS0	I	Active if SPORT 0 is enabled
TFS0	O	Driven if configured internal or in multichannel mode and SPORT 0 enabled, high impedance otherwise
RFS0	I	Active if SPORT 0 is enabled
RFS0	O	Driven if configured internal and SPORT 0 enabled, high impedance otherwise

194

System Interface   9

DR0	I	Active if SPORT 0 is enabled
DT0	O	Driven if serial port operating. Output may be static or changing depending upon serial clock, high impedance otherwise
SCLK1	I	Active
SCLK1	O	Driven to a static level if internal, high impedance otherwise
TFS1/$\overline{IRQ1}$	I	Active if SPORT 1 is enable or configured alternate ($\overline{IRQ1}$)
TFS1	O	Driven if SPORT 1 is enabled and configured for internal transmit framing, high impedance otherwise
RFS1/$\overline{IRQ0}$	I	Active if SPORT 1 is enabled or configured alternate ($\overline{IRQ0}$)
RFS1	O	Driven if SPORT 1 is enabled and configured for internal receive framing, high impedance otherwise
DR1/FLAGIN	I	Active if SPORT 1 is enabled or configured alternate (FLAGIN)
DT1/FLAGOUT	O	Driven if serial port operating. Output may be static or changing depending upon serial clock. Driven if SPORT 1 is enabled or configured alternate (FLAGOUT)
BMODE	I	Active
HSIZE	I	Active
HMD0	I	Active
HMD1	I	Active
HSEL	I	Active
HRD	I	Active
HWR	I	Active
HADR<2:0>	I	Active
HDATA<15:0>	I	Active if host writing or HMD1 and HA2/ HALE HIGH, inactive otherwise
HDATA<15:0>	O	Driven if host reading, high impedance otherwise
HACK	O	Driven
FL<2:0>	O	Driven to previous value
VIN (NORM)	I	Inactive, set analog powerdown [bit 5 of DM(0x3FEE)]
VIN (AUX)	I	Inactive, set analog powerdown bit
VFB (NORM)	O	Inactive, set analog powerdown bit
VFB (AUX)	O	Inactive, set analog powerdown bit
VOUT P	O	Driven low in powerdown
VOUT N	O	Driven low in powerdown
VREF	O	Reference turned off

Table 9.5 ADSP-21msp50 Pin Status During Powerdown

9 System Interface

To assure the lowest power consumption, all active input pins should be held at a CMOS level. All active output pins should be free of resistive load since load current will increase power dissipation. Some pins will be in one of several states depending upon the connection of mode pins. For example, the HIP data bus pins may be either active or inactive depending whether a host write is in progress or how the host mode pins are connected. You must perform a careful analysis of each input and output pin in order to insure lowest power dissipation.

Some inputs are active but ignored. The state of these inputs does not matter as long as they are at a CMOS level.

9.8.7   PWDACK Pin

A powerdown acknowledge pin (PWDACK) is provided on some ADSP-2100 family processors. The powerdown acknowledge pin is an output which indicates when the processor is powered down. It is driven high by the processor when it has powered down and is driven low when the processor has completed its powerup activity. A low level on the PWDACK pin also indicates that there is a valid CLKOUT signal and that instruction execution has begun. Figure 9.6 shows an example of timing for the powerdown and restart sequence.

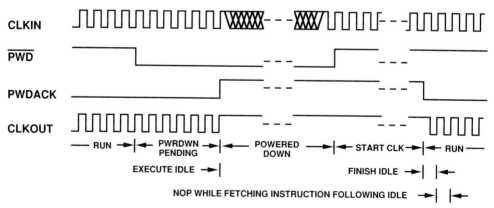

Figure 9.6 Powerdown Timing Example

System Interface   9

The processor is executing code when the $\overline{PWD}$ pin is brought low. The processor vectors to the powerdown interrupt vector and an IDLE instruction is executed causing the processor to go into powerdown. The CLKOUT and PWDACK signals are driven high by the processor. At this point, the input clock pin is ignored. If the processor is put into the powerdown mode via the powerdown force bit in the powerdown control register, the result is the same as described above.

The input clock is started and the $\overline{PWD}$ pin is brought high. After the necessary start-up cycles the processor brings the PWDACK output low, begins driving the CLKOUT pin with a clock signal and begins to fetch the instruction after the IDLE instruction. The processor then resumes normal operation.

When powerdown is terminated with the $\overline{RESET}$ pin or if a start-up delay is selected, a low level on the PWDACK pin only indicates the start of oscillations on the CLKOUT pin. It will not necessarily indicate the start of instruction execution.

The state of PWDACK and also the CLKOUT signal is undefined during the first 100 cycles of initial reset.

9.8.8   Using Powerdown As A Non-Maskable Interrupt

The powerdown interrupt is never masked. It is possible to use this interrupt for other purposes if desired. The processor will not go into powerdown until an IDLE instruction is executed. If an RTI is executed before the IDLE instruction, then the processor returns from the powerdown interrupt and the powerdown sequence is aborted.

It is possible to place a series of instructions at the powerdown interrupt vector location 0x002C. This routine should end with an RTI instruction and not contain an IDLE instruction if the interrupt is to be used for purposes other than powerdown.

Memory Interface ■ 10

10.1   OVERVIEW

The ADSP-2100 family uses a modified Harvard architecture, in which data memory stores data, and program memory stores both instructions and data. Family processors other than the ADSP-2100 processor contain on-chip RAM and/or ROM, so that a portion of the program memory space and a portion of the data memory space reside on-chip. These processors also have a boot memory space in addition to the data and program spaces. The boot memory space allows you to load program memory with code from an external source (EPROM, for example) at reset.

In each ADSP-2100 family device, memory is connected with the other functional units by four on-chip buses: the data memory address bus, data memory data bus, program memory address bus, and program memory data bus. In the ADSP-2100, all four of these buses are extended off-chip. In the other devices, which have internal memory, a single external memory address bus and and a single external data bus are extended off-chip; these buses are used for program, data, and boot memory access.

This chapter is divided into two sections: the first describes the memory interface of the ADSP-2101, ADSP-2105, ADSP-2111, and ADSP-21msp50 processors; the second describes the memory interface of the ADSP-2100. For each bus transaction, only the sequence of events is described; you must consult the data sheets for actual timing parameters.

10.2   ADSP-2101, ADSP-2105, ADSP-2111 & ADSP-21MSP50

In these ADSP-2100 family devices, which have internal memory, the internal program memory address (PMA) bus and internal data memory address (DMA) bus are multiplexed into a single address bus which is extended off-chip. Likewise, the internal program memory data (PMD) bus and internal data memory data (DMD) bus are multiplexed into a single external data bus. The sixteen MSBs of the external data bus are used as the DMD bus. In other words, D_{23-8} are used for DMD_{15-0}.

10 Memory Interface

These family processors (i.e. all except the ADSP-2100) require the use of an instruction rate clock. For example: for 12.5 MHz operation, a 12.5 MHz clock is required. All timing diagrams for these processors use CLKOUT (clock out) as reference.

There are three separate memory spaces: data memory, program memory and boot memory. The $\overline{PMS}$, $\overline{DMS}$ and $\overline{BMS}$ signals indicate which memory is being accessed. Because program memory and data memory buses are shared, if more than one off-chip transfer needs to be made in the same instruction there will be an overhead cycle required. There is no overhead if just one off-chip access with no wait states occurs in any instruction. Figure 10.1 shows the external memory buses and control signals.

All external memories may have automatic wait state generation associated with them. The number of wait states—each equal to one instruction cycle—is programmable.

10.2.1 Boot Memory Interface

The entire internal program memory, or any portion of it, can be loaded from an external source using a boot sequence. To interface with inexpensive EPROM, the processor loads instructions one byte at a time.

Automatic booting at reset depends on the state of the MMAP (memory map control signal) pin at the time of processor reset. The boot sequence occurs if the MMAP pin is logical 0. The boot sequence can also be initiated after reset by software.

Processors that include a Host Interface Port (HIP) can boot using either the memory interface or the HIP (from a host computer). The state of the BMODE pin determines which method is used (memory interface if BMODE is 0; HIP if BMODE is 1). Booting through the HIP is described in Chapter 7.

$\overline{BR}$ is recognized during the booting sequence. The bus is granted after completion of loading the current byte.

Special versions of the ADSP-2100 family processors, such as the ADSP-2102, contain on-chip program memory ROM; with these devices, no booting occurs. Refer to the data sheet of these devices for specifics.

Memory Interface 10

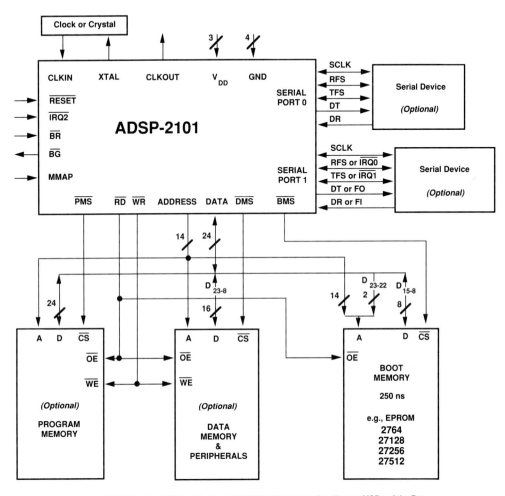

NOTE: The two MSBs of the Boot EPROM Address are also the two MSBs of the Data Bus. This is only required for the 27256 and 27512.

Figure 10.1 ADSP-2101 With External Memory

10 Memory Interface

10.2.1.1 Boot Pages

Boot memory is organized into eight pages, each of which can 8K bytes long. Every fourth byte of a page is an "empty" byte, except the first one, which contains the page length. Each set of three bytes between successive empty bytes contains an instruction. The page length is read first and then bytes are loaded from the top of the page downwards. This results in shorter booting times for shorter pages.

The length of the boot page is given as:

page length = (number of 24-bit PM words/8) – 1

That is, a page length of 0 causes the boot address generator to generate byte addresses for 8 words which reside in 32 sequential ROM locations.

The ADSP-2100 Family PROM Splitter, part of the ADSP-2100 Family Development Software tools, calculates the proper page length for your program and orders the bytes of your program as shown in Figure 10.2.

Address

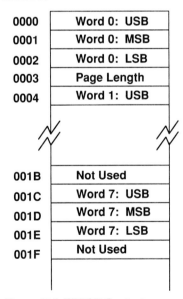

Figure 10.2 EPROM Contents

Memory Interface   10

10.2.1.2 Powerup Boot And Software Reboot

Upon a hardware or software reset, the boot sequence occurs if the MMAP pin is a logical 0. The boot sequence on reset always loads boot page 0. After reset, boot loading can occur under program control from any one of up to 8 different boot pages. The boot page select field (BPAGE) in the memory-mapped register at location 0x3FFF (see Figure 10.3) specifies which boot page is to be loaded. To boot from a specific boot page, set BPAGE to the desired page number and, in the same memory-mapped register, set the boot force bit (BFORCE). When the boot force bit is set, the (software-forced) booting sequence starts. Except for the page selection and (possibly) the number of wait states, there is no difference between a software-forced boot sequence and a reset boot sequence.

Tables in Chapter 9 show the state of the processor control registers after a reset and after a software reboot. Essentially, the processor's control state is saved, but stacks are cleared and execution starts at the restart vector, at program memory location 0.

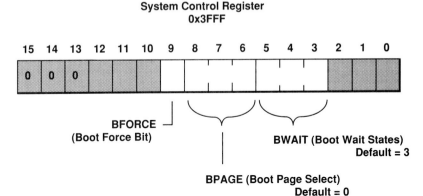

Figure 10.3 Boot Control Fields In System Control Register

10.2.1.3 Boot Memory Access

The processor can boot its internal memory from a single byte-wide CMOS EPROM, such as the 27C64 and 27C512. A low-cost, commodity-grade EPROM with an industry-standard access time can be used. The number of wait states for the boot memory access is located in the BWAIT field of the memory-mapped register located at address 0x3FFF (Figure 10.3). This field can be set to any value from 0 to 7 in order to generate 0 to 7 wait states. The default value at reset is three wait states.

203

10 Memory Interface

Timing of the boot memory access is identical to that of external program memory or external data memory accesses, except that the active strobe is $\overline{BMS}$ rather than $\overline{PMS}$ or $\overline{DMS}$. To address eight pages of 8K bytes each, 16 bits are needed. The least significant 14 bits are output on the 14-bit address bus, and the most significant 2 bits are output on the 2 MSBs of the data bus during a boot memory access. Data is read from the data bus on the middle eight bits.

10.2.1.4 Boot Loading Sequence

The order in which the processor loads data into its internal memory during a boot operation is unimportant in most applications. The boot loading sequence is explained in this section for those instances in which the order is relevant, for instance when a latch is providing data rather than an EPROM.

To execute the boot operation, the boot address generator generates the appropriate byte addresses and loads the internal program memory with the contents of the EPROM. The internal program memory is loaded beginning with the high addresses. For example, assume that eight 24-bit words are loaded into the processor during the booting process. The first word written into program memory is written to address 0007. The last word loaded is written to internal program memory address 0000.

The boot address is made up of several values, as shown in Figure 10.4: the 3-bit page number (from BPAGE in the system control register); the 8-bit page length, which is always read first, from the fourth byte of the page; three ones (111); and a 2-bit code whose value determines which byte of the word is being addressed.

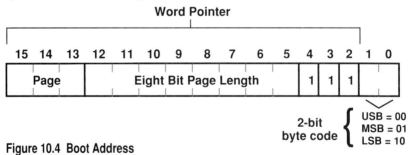

Figure 10.4 Boot Address

The last 24-bit word (the last instruction or program memory data value) is loaded into the processor first. The byte loading order is: upper byte, lower byte, middle byte. The word pointer is then decremented. This addresses the second-to-last 24-bit word in the EPROM.

Memory Interface 10

For example, to boot from page 0 the shortest allowable page (eight 24-bit words corresponding to a page length of 0), the following addresses would be generated:

1. The first address generated is 0003 which reads the page length.

2. The next address generated in this example is address 001C. This is the upper byte of the last word.

3. The byte code is then updated to specify the lower byte (the final two bits are 10) and the address generated is 001E.

4. The byte address changes again, this time to address the middle byte (the two bit code is 01) and the address generated is 001D.

5. Once all three bytes are loaded, the word counter is decremented. The three succeeding byte addresses generated are 0018, 001A, and 0019.

6. The word counter is decremented again and the next set of byte addresses generated is 0014, 0016, and 0015. This process continues until word 0 is loaded.

The contents of the EPROM, the byte addresses and the order of addresses generated is summarized in Figure 10.5, which can be found on the following page.

10.2.2 Program Memory Interface

The processors address 16K of 24-bit wide program memory, up to 2K on-chip and the remainder external, using the control lines shown in Figure 10.1. The processors supply a 14-bit address on the program memory address bus (PMA) which is driven off-chip on the address bus in the case of external program memory accesses. Instructions or data are transferred across the 24-bit program memory data (PMD) bus which is also multiplexed off-chip. For a dual off-chip data fetch, the data from program memory is read first, then the data memory data. A program memory select pin, $\overline{PMS}$, indicates that the address bus is being driven with a program memory address and memory can be selected.

Two control lines indicate the direction of the transfer. Memory read ($\overline{RD}$) is active low signaling a read and memory write ($\overline{WR}$) is active low for a write operation. Typically, you would connect $\overline{PMS}$ to $\overline{CE}$ (Chip Enable), $\overline{RD}$ to $\overline{OE}$ (Output Enable) and $\overline{WR}$ to $\overline{WE}$ (Write Enable) of your memory.

10 Memory Interface

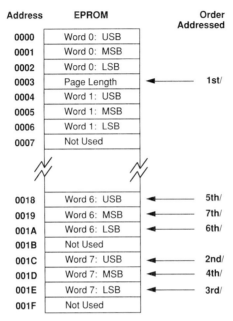

Figure 10.5 Boot Loading Order

10.2.2.1 Program Memory Read / Write

On-chip memory accesses do not drive any external signals. $\overline{PMS}$, $\overline{DMS}$, $\overline{RD}$, and $\overline{WR}$ remain high (deasserted); the address and data buses are tristated. Off-chip program memory access happens in this sequence:

1. The processor places the address on the PMA bus, which is multiplexed off-chip, and $\overline{PMS}$ is asserted.

2. $\overline{RD}$ or $\overline{WR}$ is asserted.

3. Within a specified time, data is placed on the data bus, multiplexed to the internal PMD bus.

4. The data is read or written and $\overline{RD}$ (or $\overline{WR}$) is deasserted.

5. $\overline{PMS}$ is deasserted.

The basic read and write cycles are illustrated in Figure 10.6. Part A shows zero wait states and Part B shows the effect of one wait state.

Memory Interface   10

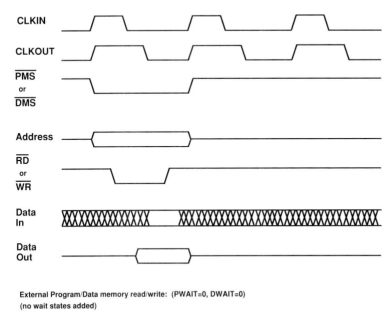

External Program/Data memory read/write: (PWAIT=0, DWAIT=0)
(no wait states added)

Figure 10.6A Memory Read And Write, No Wait States

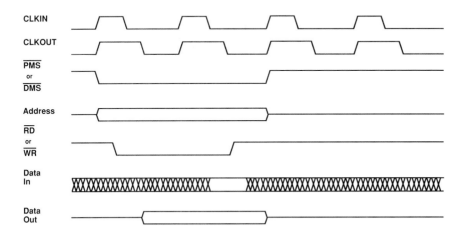

External Program/Data memory read/write: (PWAIT=1, DWAIT=1)

(one wait state added due to WAIT register)

Figure 10.6B Memory Read And Write, One Wait State

10 Memory Interface

External program memory has a programmable wait state field (PWAIT) in the system control register, as shown in Figure 10.7. PWAIT defaults to seven wait states for program memory access on power-up.

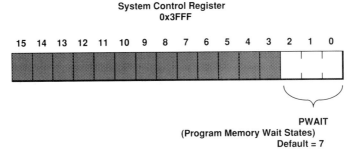

Figure 10.7 Program Memory Wait State Field In System Control Register

10.2.2.2 Program Memory Map

Depending on the state of the MMAP pin, the program memory space is mapped in one of the configurations shown in Figure 10.8.

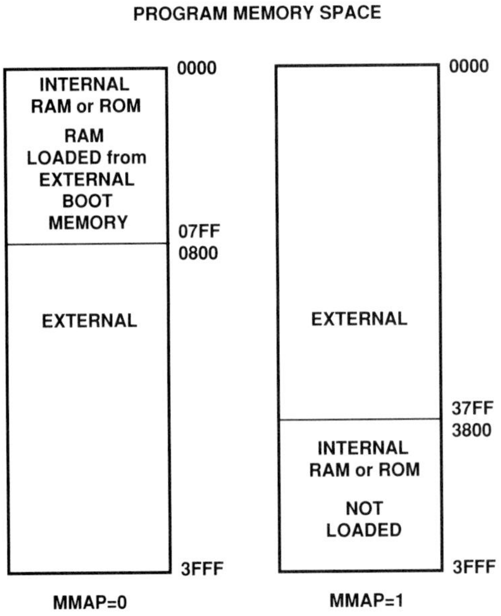

PROGRAM MEMORY SPACE

Figure 10.8 Program Memory Configurations

Memory Interface   10

The 16K program memory space can hold instructions and data intermixed in any combination. The Linker determines where to place relocatable code and data segments. You may specify absolute address placement for any module or data structure, including the code for the restart and interrupt vector locations. The restart vector is at program memory address 0000. The locations of interrupt vectors are given in Table 3.2 of Chapter 3.

Internal program memory RAM is fast enough to supply an instruction and data in the same cycle eliminating the need for cache memory as in the ADSP-2100. Consequently, if the processor is operating entirely from on-chip memories, it can fetch two operands and the next instruction on every cycle. It can also fetch any one of these three from external memory with no performance penalty.

The ADSP-21msp51 processor contains an additional 2K words of mask-programmable ROM. The ROM always resides at program memory locations 0800 through 09FF. The ROM is enabled by setting the ROMENABLE bit of the Data Memory Wait State control register. With this bit set, addressing program memory in this range will access the on-chip ROM, as shown in Figure 10.9.

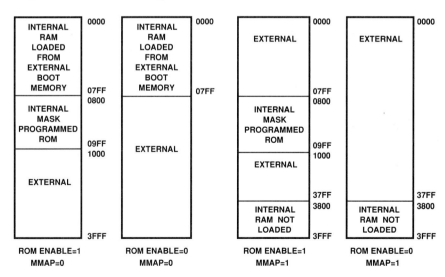

Figure 10.9 ADSP-21msp51 Program Memory Configurations (ROM Enabled)

10 Memory Interface

10.2.3 Data Memory Interface

The processor can address 16K words of 16-bit data memory as shown in Figure 10.1. On-chip data memory is up to 1K in size, and begins at 0x3800 for the ADSP-2101, ADSP-2102, ADSP-2111, ADSP-2112, ADSP-21msp50 and ADSP-21msp51. On-chip data memory is 512 locations for the ADSP-2105 and the ADSP-2106, beginning at 0x3800. The control registers are mapped into the top 1K of data memory, that is 0x3C00-0x3FFF in all processors with on-chip memory; the rest of the top 1K is reserved for future expansion. 14K of external data memory is available for user data storage.

The processor supplies a 14-bit address on the data memory address bus (DMA) which is multiplexed off-chip. Data is transferred across the upper 16 bits of the 24-bit memory data bus, which is also multiplexed off-chip. A data memory select pin, $\overline{DMS}$, indicates that the address bus is being driven with a data memory address and memory can be selected.

Two control lines indicate the direction of the transfer. Memory read ($\overline{RD}$) is active low signaling a read and memory write ($\overline{WR}$) is active low for a write operation. Typically, you would connect $\overline{DMS}$ to $\overline{CE}$ (Chip Enable), $\overline{RD}$ to $\overline{OE}$ (Output Enable) and $\overline{WR}$ to $\overline{WE}$ (Write Enable) of your memory.

10.2.3.1 Data Memory Read/Write

The on-chip data memory access is transparent to the outside memory interface. Only off-chip accesses drive the memory interface. Off-chip data memory access requires the same sequence as for off-chip program data memory, namely:

1. The processor places the address on the DMA bus, which is multiplexed off-chip, and $\overline{DMS}$ is asserted.

2. $\overline{RD}$ or $\overline{WR}$ is asserted.

3. Within a specified time, data is placed on the data bus, multiplexed to the internal DMD bus.

4. The data is read or written and $\overline{RD}$ (or $\overline{WR}$) is deasserted.

5. $\overline{DMS}$ is deasserted.

The basic read and write cycles are illustrated in Figure 10.6 in the preceding section.

210

Memory Interface 10

10.2.3.2 Data Memory Map

Data memory configurations are shown in Figure 10.10. Each of the five zones of off-chip data memory has its own programmable wait state. Wait states are extra cycles that the processor either waits before latching data (on a read) or drives the data (on a write). This means that one zone of memory could be used for working with memory-mapped peripherals of one speed while another zone was used with faster or slower peripherals. Similarly, slower and faster memories can be used for different purposes, as long as they are located in different zones of the data memory map.

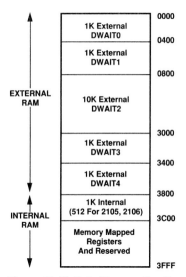

2101 DATA MEMORY SPACE

Figure 10.10 Data Memory Configurations

The data memory wait state control register, shown in Figure 10.11 on the following page, has a separate field for each zone of external memory. Each 3-bit field contains the number (0-7) of wait states for the corresponding zone of memory; the default value in each case is 7. In the ADSP-21msp51 processor, one bit in this register is used to enable or disable the on-chip ROM (see "Program Memory Map").

10 Memory Interface

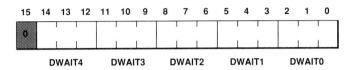

Data Memory Wait State Control Register
0x3FFE

Figure 10.11 Data Memory Wait State Control Register

10.2.3.3 Parallel And Memory-Mapped Peripherals

Peripherals requiring parallel communications and other types of devices can be mapped into external data memory. Communication takes the form of reading and writing the memory locations associated with the device. Some A/D and D/A converters require this type of interface. The PORT directives in the System Builder and Assembler modules of the ADSP-2100 Family Development Software support this mapping. Communication with a memory-mapped device consists simply of reading and writing the appropriate locations. By matching the access times of the external devices to the wait states specified for their zone of data memory, you can easily interface a variety of devices.

10.2.4 Bus Request / Grant

Using the bus request signal, $\overline{BR}$, and the bus grant signal, $\overline{BG}$, the processor can relinquish control of the external memory interface, giving access to an external device such as a host processor. If the GO mode is enabled, the processor continues to execute instructions using on-chip program and data memory. The processor halts only when it must access external memory. If the GO mode is not enabled, the processor always halts before granting the bus. The serial ports and host interface port remain active during a bus grant, whether or not the processor core halts.

The external device requests the bus by asserting $\overline{BR}$. $\overline{BR}$ is a synchronous input with setup and hold requirements which must be met, as specified in the individual device data sheets. Because $\overline{BR}$ is synchronous, the processor is able to respond to the request in the same cycle. This feature is a benefit of synchronous systems. If the system generates $\overline{BR}$ as an asynchronous signal, a simple synchronization circuit (such as that shown in Figure 10.12) must be used to eliminate any synchronization delays.

When $\overline{BR}$ is recognized, the processor responds immediately by asserting $\overline{BG}$ during the same cycle, unless a memory access is in progress. In this case, the memory access will be completed before $\overline{BG}$ is asserted. The

Memory Interface 10

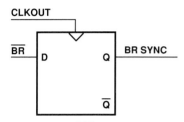

Figure 10.12 $\overline{BR}$ Synchronization Circuit

processor halts if necessary and tristates the address bus lines, the data bus lines, $\overline{WR}$, $\overline{RD}$, $\overline{PMS}$, $\overline{DMS}$ and $\overline{BMS}$. Control is then transferred to the requesting device by asserting $\overline{BG}$. The sequence of events is illustrated in Figure 10.13.

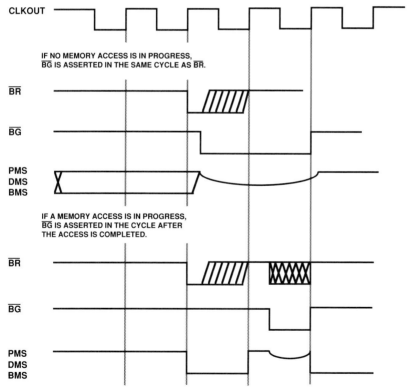

Figure 10.13 Bus Request With And Without Memory Access

10 Memory Interface

If the processor is in the middle of an instruction requiring the access of both external program and external data memory (requiring two consecutive cycles of external bus use) and the second access has not yet begun, BG is granted in between the two accesses. The second access is performed after the bus request is removed.

Even if the processor has to halt, its internal state is not affected by granting the bus. After the bus request is released by the external device, normal operation resumes from the point at which it was halted. This applies uniformly to all processor operations.

The external device returns control by deasserting BR. The processor responds by deasserting BG and resuming control of the bus. BG is always deasserted in the same cycle that the removal of bus request is recognized. Refer to the device data sheet for exact timing relationships.

Bus request can be asserted while the processor is in reset. During reset, BG is always asserted in the same cycle that BR is recognized. BR is also recognized during the booting sequence. The bus is granted after completion of loading the current byte, including any wait states. Using BR during a boot operation is one way to bring booting under the control of a host computer. See Chapter 14 for an example of how this is done.

10.2.5 Memory Interface Summary
Table 10.1 summarizes the states of the memory interface pins for various combinations of program memory and data memory accesses. Table 10.2 summarizes the states of the memory interface and control pins during reset, booting and bus grant.

10.3 ADSP-2100
The memory interface supports transfers between the ADSP-2100 and program and data memories using the control lines shown in Figure 10.14, which can be found on page 216. The ADSP-2100 is clocked from a clock signal that is 4 times the instruction rate. The ADSP-2100 features an off-chip Harvard architecture. There are separate buses for interfacing to data memory and for program memory. The program memory interface provides for the synchronous transfer of instructions and data between the program memory and the processor. The data memory interface provides for the synchronous transfer between data memory and the processor.

Memory Interface   10

Access	$\overline{PMS}$	$\overline{DMS}$	$\overline{BMS}$	$\overline{RD}$	$\overline{WR}$	Address	Data
Internal program memory only	high	high	high	high	high	tristated*	tristated
Internal data memory only	high	high	high	high	high	tristated	tristated
Internal program memory, external data memory	high	low	high	low (for read)	low (for write)	DM address	DM data
Internal data memory, external program memory	low	high	high	low (for read)	low (for write)	PM address	PM data
External boot memory	high	high	low	low (for read)	high	Boot address	Boot data, Boot page address

* ADSP-2101 Emulator does not tristate the address bus.

Table 10.1 Pin States During Memory Accesses

Operation	Address	Data	$\overline{PMS}$ $\overline{DMS}$ $\overline{BMS}$	$\overline{RD}$ $\overline{WR}$	CLKOUT	SPORTs FO	$\overline{BG}$
Reset	tristated	tristated	high	high	active	tristated	high
Auto Booting after Reset	active	active	$\overline{BMS}$ active $\overline{PMS}$, $\overline{DMS}$ high	$\overline{RD}$ active $\overline{WR}$ high	active	tristated	high
$\overline{BR}$ Asserted during Normal operation, Booting or Go Mode	tristated	tristated	tristated	tristated	active	active	low
$\overline{BR}$ Asserted during Reset	tristated	tristated	tristated	tristated	active	tristated	low

Table 10.2 Pin States During Reset, Booting, And Bus Grant

10 Memory Interface

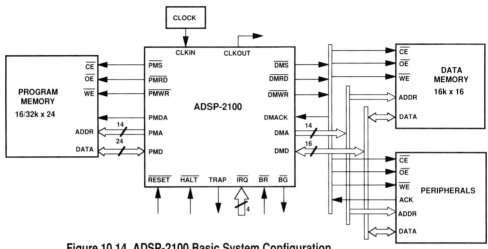

Figure 10.14 ADSP-2100 Basic System Configuration

10.3.1 Clock Signals & Processor States

The ADSP-2100 has two clock signals, CLKIN and CLKOUT. CLKIN is a master input clock to the processor that operates at four times the instruction cycle rate. The phases of CLKIN are used to define eight (1-8) distinct time periods, called the processor states, that make up an instruction cycle. This is shown in Figure 10.15. The eighth state of each instruction cycle is a dead state that provides a neutral halting point for the processor when operation is suspended. All timing diagrams annotate the phases of CKLIN with these state numbers.

CLKOUT is an output clock from the ADSP-2100 that operates at the instruction cycle rate. It is produced by dividing the frequency of CLKIN by four. The phase of CLKOUT is such that it allows external synchronization to the internal states of the processor. The falling transition of CLKOUT always occurs at the transition between states three and four while the rising edge always occurs at the transition between states seven and eight. This relationship is shown in Figure 10.15.

10.3.2 Synchronization Delay

The ADSP-2100 has several asynchronous inputs, namely, $\overline{\text{RESET}}$, $\overline{\text{HALT}}$, $\overline{\text{BR}}$, DMACK and $\overline{\text{IRQ0-3}}$. These inputs can be asserted in arbitrary phase to the processor clock, CLKIN. The ADSP-2100 resynchronizes them prior to recognizing them. The delay associated with resynchronization and eventual recognition is called the synchronization delay.

Memory Interface 10

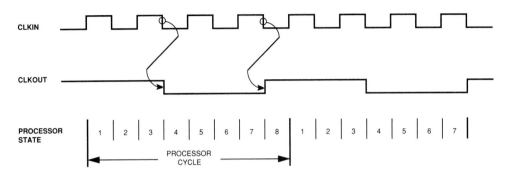

Figure 10.15 Clock Signals & Processor States

Different asynchronous inputs are recognized at different points in the processor cycle. For example, $\overline{\text{HALT}}$ is recognized at the end of state three but interrupt requests are recognized at the end of state seven.

Any asynchronous input must be valid prior to the recognition point. The minimum time prior to recognition (the setup time) is given on the data sheet. If an input does not meet the setup time on a given cycle, it will be recognized during the next cycle if it is held valid. Therefore, to ensure recognition of an asynchronous input, it must be asserted for at least one full processor cycle.

10.3.3   Program Memory Interface

The processor supplies a 14-bit address on the program memory address (PMA) bus. Data or instructions are then transferred across a 24-bit program memory data (PMD) bus. A Program Memory Select pin, $\overline{\text{PMS}}$, indicates that the address bus is being driven and memory can be selected. $\overline{\text{PMS}}$ is asserted on processor cycles in which instructions or data are fetched from program memory. Since the program memory bus of the ADSP-2100 is active in every instruction cycle for either an opcode or a data fetch, in practice $\overline{\text{PMS}}$ is asserted continuously. The only exceptions are during $\overline{\text{HALT}}$, TRAP or when the bus is tristated. $\overline{\text{PMS}}$ may be connected to the chip select on memories.

Two control lines determine the direction of the transfer. Program Memory Read, $\overline{\text{PMRD}}$, is active low indicating a memory read. $\overline{\text{PMRD}}$ is timed so that it may be used as an output enable signal. Program Memory Write, $\overline{\text{PMWR}}$, corresponds to a memory write. $\overline{\text{PMWR}}$ is timed so that it may be used as a write strobe.

217

10 Memory Interface

The program memory can be used to store both instructions and data. The processor distinguishes between these two by asserting PMDA (Program Memory Data Access) during a data transfer. The timing of PMDA is similar to the PMA bus, allowing PMDA to be used as an additional address bit. When used as the most significant address bit, the processor can address 32K words of program memory of which 16K is dedicated to data storage. Systems requiring less than 16K of program memory may allocate storage to mixed instructions and data without restriction. The System Builder module of the ADSP-2100 Family Development Software allows you to define memory use in software during development. The Development Software system uses this definition of code and data memory allocation in the Simulator, Linker and PROM Splitter.

10.3.3.1 Program Memory Read Cycle
Program memory reads occur across the program memory interface in the sequence listed below. This list gives only the functional sequence of events; for exact timing information, consult the data sheet.

1. The ADSP-2100 places the address on the PMA bus, asserts $\overline{PMS}$, drives PMDA high or low, then asserts $\overline{PMRD}$. $\overline{PMRD}$ may be used as an output enable signal. $\overline{PMS}$ remains asserted without a change of state if it was asserted on the previous cycle.

2. Within a specified time period (see data sheet), valid data must be placed on the PMD bus by the memory.

3. The ADSP-2100 reads the data on the PMD bus. (The data is strobed in on the rising edge of $\overline{PMRD}$.)

4. The ADSP-2100 de-asserts $\overline{PMRD}$ and terminates the cycle.

A timing diagram for this operation is shown in Figure 10.16.

10.3.3.2 Program Memory Write Cycle
Program memory is written with the following sequence of operations:

1. The ADSP-2100 places the address on the PMA bus, asserts $\overline{PMS}$, and drives PMDA; $\overline{PMS}$ remains asserted if already asserted from the previous cycle.

2. The ADSP-2100 places data on the PMD bus and asserts $\overline{PMWR}$ for writing.

218

Memory Interface 10

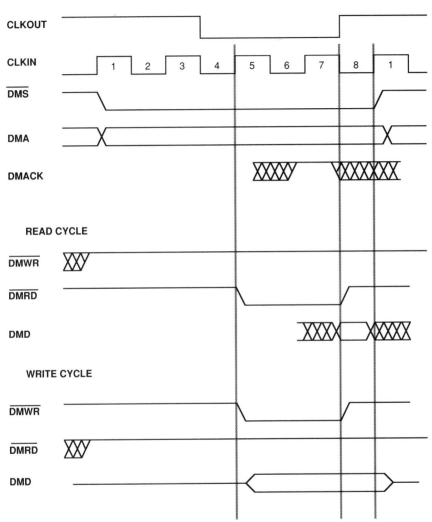

Figure 10.17 ADSP-2100 Data Memory Read/Write

10 Memory Interface

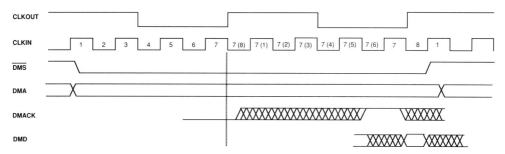

Figure 10.18 ADSP-2100 Data Memory Read Extended By DMACK

10.3.4.1 Data Memory Read Cycle
Data memory reads occur across the data memory interface as shown in Figure 10.19.

A timing diagram for this operation is shown in Figure 10.17.

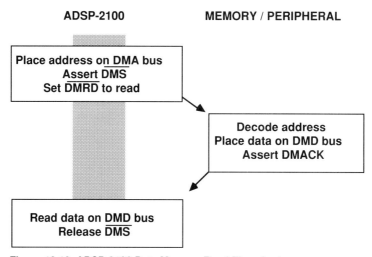

Figure 10.19 ADSP-2100 Data Memory Read Flowchart

222

10.3.4.2 Data Memory Write Cycle

Data memory is written with the sequence of operations shown in Figure 10.20.

Figure 10.17 depicts the timing for the data memory write operation. Note that $\overline{\text{DMWR}}$ may be used as a write strobe.

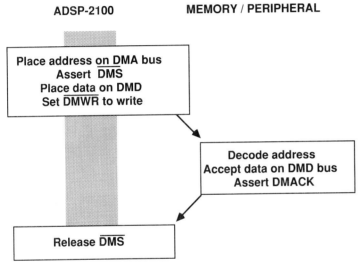

ADSP-2100   **MEMORY / PERIPHERAL**

Place address on DMA bus
Assert $\overline{\text{DMS}}$
Place data on DMD
Set $\overline{\text{DMWR}}$ to write

Decode address
Accept data on DMD bus
Assert DMACK

Release $\overline{\text{DMS}}$

Figure 10.20 ADSP-2100 Data Memory Write Flowchart

10.3.5   Bus Request / Grant

Using the bus request, $\overline{\text{BR}}$, and bus grant, $\overline{\text{BG}}$, signals, the ADSP-2100 can relinquish control of both the program and data memory interface giving direct memory access to an external device, such as a host computer.

The external device requests the bus by asserting $\overline{\text{BR}}$ (bus request). $\overline{\text{BR}}$ is recognized at the end of the next state three and the ADSP-2100 halts in state eight of that instruction cycle. $\overline{\text{BG}}$ (bus grant) is asserted at the end of state three of what would have been the next instruction cycle, i.e. four cycles of CLKIN after the bus request is recognized. This is the normal synchronous mode of servicing this request.

10 Memory Interface

The ADSP-2100 tristates all the bus driving lines: PMA, PMD, PMDA, PMWR, PMRD and PMS on the program memory interface and DMA, DMD, DMWR, DMRD and DMS on the data memory interface. Control of the memory is thus transferred to the requesting device.

The ADSP-2100's internal state is not affected by this operation. After BR is de-asserted (i.e., after the external device gives up control of the memory), normal operation resumes from the point at which it was halted. This applies uniformly to all processor operations, including the extra cycle inserted by the processor when a program memory data access is performed and the cache contents are not valid. BR can be serviced between the two cycles of a multiple cycle operation.

The device returns control to the processor by releasing BR. Four cycles of CLKIN after BR is recognized as released, the processor releases BG and takes over the bus, resuming with state one of the next cycle. Figure 10.21 is an operations flowchart. Figure 10.22 shows the relative timing of this cycle.

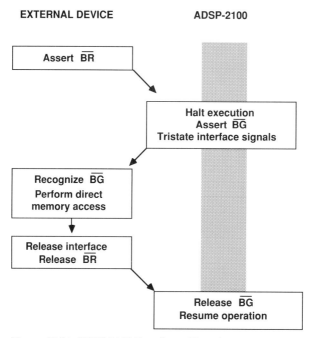

Figure 10.21 ADSP-2100 Bus Grant Flowchart

Memory Interface   10

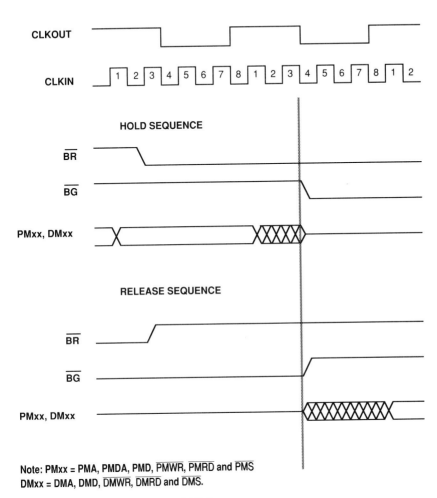

Note: PMxx = PMA, PMDA, PMD, $\overline{PMWR}$, $\overline{PMRD}$ and $\overline{PMS}$.
DMxx = DMA, DMD, $\overline{DMWR}$, $\overline{DMRD}$ and $\overline{DMS}$.

Figure 10.22 ADSP-2100 Bus Hold/Release

A bus request can be made, i.e. $\overline{BR}$ may be asserted, during a reset of the
ADSP-2100. The timing is different than shown in Figure 10.22. In this
case, $\overline{BG}$ will be asserted asynchronously, after $\overline{BR}$ is recognized. The
delay is solely due to propagation delay and is much shorter than the
synchronization delay seen during normal operation. Releasing $\overline{BR}$ causes
$\overline{BG}$ to be deasserted asynchronously.

10 Memory Interface

$\overline{BR}$ must be removed before or coincident with the removal of $\overline{RESET}$ to ensure proper operation of the processor. In other words, $\overline{BR}$ can be asserted "during" $\overline{RESET}$ but $\overline{RESET}$ should not be asserted "during" (i.e. ending before) a $\overline{BR}$.

If the bus is requested during $\overline{HALT}$ or TRAP, the request is latched and serviced after the normal synchronization delay. The processor remains halted, but tristates the busses.

Programming Model ■ 11

11.1 OVERVIEW

From a programming standpoint, a processor in the ADSP-2100 family
consists of three computational units, two data address generators, and a
program sequencer, and may also have on-chip peripherals and/or
memory, the kind and amount of which varies with each processor.
Almost all operations using these architectural components involve one or
more registers—to store data, to keep track of values such as pointers, or
to specify operating modes, for example.

Internal registers hold data, addresses, control information or status
information. For example, AX0 stores an ALU operand (data); I4 stores a
DAG2 pointer (address); ASTAT contains status flags from arithmetic
operations; and fields in DWAIT control the numbers of wait states for
different zones of data memory.

There are two types of accesses for registers. Dedicated registers such as
MX0 and IMASK can be read and written explicitly in assembly language.
For example:

```
MX0=1234;
IMASK=0xF;
```

Memory-mapped registers—the system control register, DWAIT register,
timer registers, SPORT registers and analog control registers—are
accessed by reading and writing data memory locations. For example, this
code clears the DWAIT register, which is mapped to data memory location
0x3FFE:

```
AX0=0;
DM(0x3FFE)=AX0;
```

(AX0 is used to hold the constant 0 because there is no instruction to write
an immediate data value to memory using an immediate address.)

11 Programming Model

All of the registers available in the ADSP-2100 family are shown in Figure 11.1. Not all of these registers are available on every member of the family. The registers are grouped by function: data address generators (DAGs), program sequencer, computational units (ALU, MAC and shifter), bus exchange, memory interface, timer, SPORTs, host interface and analog interface. Note that the system control register is associated with several functions.

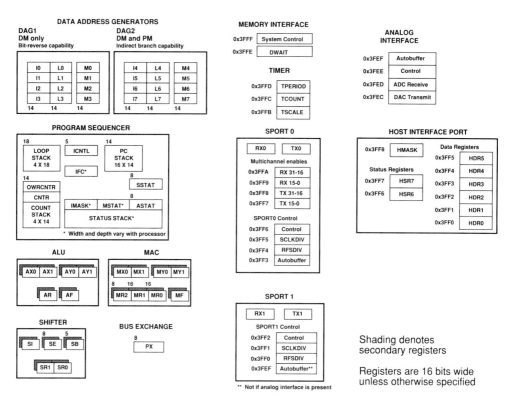

Figure 11.1 ADSP-2100 Family Registers

Programming Model 11

11.1.1 Data Address Generators

DAG1 and DAG2 each have twelve 14-bit registers: four index (I) registers for storing pointers, four modify (M) registers for updating pointers and four length (L) registers for implementing circular buffers. DAG1 addresses data memory only and has the capability of bit-reversing its outputs. DAG2 addresses both program and data memory and can provide addresses for indirect branching (jumps and calls) as well as for accessing data.

For example:

```
AX0=DM(I0,M0);
```

is an indirect data memory read from the location pointed to by I0. Once the read is complete, I0 is updated by M0.

```
PM(I4,M5)=MR1;
```

is an indirect program memory data write to the address pointed to by I4 with a post modify by M5. The instruction

```
JUMP (I4);
```

is an example of an indirect jump.

11.1.2 Program Sequencer

Registers associated with the program sequencer control subroutines, loops, and interrupts. They also indicate status and select modes of operation.

11.1.2.1 Interrupts

The ICNTL register controls interrupt nesting and external interrupt sensitivity; the IFC register lets you force and clear interrupts in software (not available on the ADSP-2100); the IMASK register masks (disables) individual interrupts. The widths of the IFC and IMASK registers depend on the processor, since different members of the ADSP-2100 family support different numbers of interrupts.

11 Programming Model

11.1.2.2 Loop Counts

The CNTR register stores the count value for the currently executing loop. The count stack allows the nesting of count-based loops to four levels. A write to CNTR pushes the current value onto the count stack before writing the new value. For example:

```
CNTR=10;
```

pushes the current value of CNTR on the count stack and then loads CNTR with 10.

OWRCNTR is a special instruction (not available on the ADSP-2100) through which you can overwrite the count value for the current loop without pushing CNTR on the count stack.

11.1.2.3 Status And Mode Bits

The stack status (SSTAT) register contains full and empty flags for stacks. The arithmetic status (ASTAT) register contains status flags for the computational units. The mode status (MSTAT) register contains control bits for various options. On the ADSP-2100 processor, MSTAT holds 4 bits that control alternate register selection for the computational units, bit-reverse mode for DAG1, and overflow latch and saturation modes for the ALU. On other processors in the family, MSTAT has an additional 3 bits to control the MAC result placement, timer enable and Go mode.

Use the Mode Control instruction (ENA, DIS) to conveniently enable or disable processor modes.

11.1.2.4 Stacks

The program sequencer contains four stacks that allow loop, subroutine and interrupt nesting.

The PC stack is 14 bits wide and 16 locations deep. It stores return addresses for subroutines and interrupt service routines, and top-of-loop addresses for loops. PC stack handling is automatic for subroutine calls and interrupt handling. In addition, the PC stack can be manually pushed or popped using the PC Stack Control instructions (PUSH, POP).

The loop stack is 18 bits wide, 14 bits for the end-of-loop address and 4 bits for the termination condition code. The loop stack is four locations deep. It is automatically pushed during the execution of a DO UNTIL instruction. It is popped automatically during a loop exit if the loop was nested. The loop stack may be manually popped with the POP LOOP instruction.

Programming Model 11

The status stack, which is automatically pushed when the processor services an interrupt, accommodates the interrupt mask (IMASK), mode status (MSTAT) and arithmetic status (ASTAT) registers. The depth and width of the status stack varies with each processor, since different processors have different numbers of interrupts. The status stack is automatically popped when the return from interrupt (RTI instruction) is executed. The status stack can be pushed and popped manually with the PUSH STS and POP STS instructions.

The count stack is 14 bits wide and holds counter (CNTR) values for nested counter-based loops. This stack is pushed automatically with the current CNTR value when there is a write to CNTR. The counter stack may be manually popped with the POP CNTR instruction.

11.1.3   Computational Units

The registers in the computational units store data.

The ALU and MAC require two inputs for most operations. The AX0, AX1, MX0 and MX1 registers store X inputs, and the AY0, AY1, MY0 and MY1 registers store Y inputs.

The AR and AF registers store ALU results; AF can be fed back to the ALU Y input, whereas AR can provide the X input of any computational unit. Likewise, the MR0, MR1, MR2 and MF register store MAC results and can be fed back for other computations. The 16-bit MR0 and MR1 registers together with the 8-bit MR2 register can store a 40-bit multipy/accumulate result.

The shifter can receive input from the ALU or MAC, from its own result registers, or from a dedicated shifter input (SI) register. It can store a 32-bit result in the SR0 and SR1 registers. The SB register stores the block exponent for block floating-point operations. The SE register holds the shift value for normalize and denormalize operations.

Registers in the computational units have secondary registers, shown in Figure 11.1 as second set of registers behind the first set. Secondary registers are useful for single-cycle context switches. The selection of these secondary registers is controlled by a bit in the MSTAT (mode status) register; the bit is set and cleared by these instructions:

```
ENA SEC_REG;      {select secondary registers}
DIS SEC_REG;      {select primary registers}
```

11 Programming Model

11.1.4 Bus Exchange

The PX register is an 8-bit register that allows data transfers between the 16-bit DMD bus and the 24-bit PMD bus. In a transfer between program memory and a 16-bit register, PX provides or receives the lower eight bits.

11.1.5 Timer

The TPERIOD, TCOUNT and TSCALE hold the timer period, count and scale factor values, respectively. These registers are memory-mapped at locations 0x3FFD, 0x3FFC, and 0x3FFB respectively.

11.1.6 Serial Ports

SPORT0 and SPORT1 each have receive (RX), transmit (TX) and control registers. The control registers are memory-mapped registers at locations 0x3FEF–0x3FFA in data memory. SPORT0 also has registers for controlling its multichannel functions. Each SPORT control register contains bits that control frame synchronization, companding, word length and, in SPORT0, multichannel options. The SCLKDIV register for each SPORT determines the frequency of the internally generated serial clock, and the RFSDIV register determines the frequency of the internally generated receive frame sync signal for each SPORT. The autobuffer registers control autobuffering in each SPORT. Note that on processors with an analog interface, SPORT1 does not have autobuffering capability (analog autobuffering is supported instead).

Programming a SPORT consists of writing its control register and, depending on the modes selected, its SCLKDIV and/or RFSDIV registers as well. The following example programs SPORT0 for 8-bit μ-law companding, normal framing, and an internally generated serial clock. RFSDIV is set to 255, for 256 SCLK cycles between RFS assertions. SCLKDIV is set to 2, resulting in an SCLK frequency that is 1/6 of the CLKOUT frequency.

```
SI=0xB27;
DM(0X3FF6)=SI;     {SPORT0 control register}

SI=2;
DM(0x3FF5)=SI;     {SCLKDIV = 2}

SI=255;
DM(0x3FF4)=SI;     {RFSDIV = 255}
```

Programming Model   11

11.1.7   Memory Interface And SPORT Enables

The system control register, memory-mapped at data memory location 0x3FFF, contains SPORT enables as well as SPORT1 configuration. It also contains fields for controlling the boot operation: selecting the page, specifying the number of wait states and forcing the boot in software. And, it contains a field that specifies the number of wait states for program memory accesses.

The DWAIT register, memory-mapped at data memory location 0x3FFE, contains fields that specify the number of wait states for each bank of data memory. In processors with optional ROM (not always present), it also contains a bit for enabling the ROM.

11.1.8   Host Interface

The host interface has six data registers, two status registers and an interrupt mask register. These registers are memory-mapped at data memory locations 0x3FE7 – 0x3FE0. The status registers contains status flags for each of the data registers. The HMASK register lets you enable or disable the generation of HIP read or HIP write interrupts independently for each HIP data register. HMASK is memory-mapped at data memory location 0x3FE8 in data memory.

11.1.9   Analog Interface

The analog interface has four memory-mapped registers. These registers are memory-mapped in data memory locations 0x3FEC – 0x3FEF. The transmit register sends data to the DAC for transmitting. The receive register receives data from the ADC. The analog control register contains bits that select amplifier, gain, analog input and filter options. It also has a bit for forcing a powerdown of the analog circuitry. The analog autobuffer/powerdown register controls autobuffering for both receiving and transmitting. It also controls the powerdown modes of the processors that have these modes.

Instruction Set Reference ■ 12

12.1　QUICK LIST OF INSTRUCTIONS

This chapter is a complete reference for the instruction set of the ADSP-2100 family. The instruction set is organized by instruction group and, within each group, by individual instruction. The list below shows all of the instructions and the reference page for each.

12 Instruction Set Reference

12.2 OVERVIEW

This chapter provides an overview and detailed reference for the instruction set of the ADSP-2100 family of DSP microprocessors. An example program is included at the end of this introduction (Listing 12.1).

For information regarding the ADSP-2100 Family Development Software, refer to the *ADSP-2100 Family Assembler Manual*, *ADSP-2100 Family Simulator Manual*, and *ADSP-2100 Family C Compiler & Runtime Library Manual*. These manuals provide a complete guide to the development software. The handbook *Digital Signal Processing Applications Using The ADSP-2100 Family*, *Volume 1*, presents ADSP-2100 family applications programs with source code and discussion.

The instruction set is tailored to the computation-intensive algorithms common in DSP applications. For example, sustained single-cycle multiplication/accumulation operations are possible. The instruction set provides full control of the processors' three computational units: the ALU, MAC and Shifter. Arithmetic instructions can process single-precision 16-bit operands directly; provisions for multiprecision operations are available.

The high-level syntax of ADSP-2100 family source code is both readable and efficient. Unlike many assembly languages, the ADSP-2100 family instruction set uses an algebraic notation for arithmetic operations and for data moves, resulting in highly readable source code. There is no performance penalty for this; each program statement assembles into one 24-bit instruction which executes in a single cycle. There are no multicycle instructions in the instruction set. (If memory access times require, or contention for off-chip memory occurs, overhead cycles will be required, but all instructions can otherwise execute in a single cycle.)

In addition to JUMP and CALL, the instruction set's control instructions support conditional execution of most calculations and a DO UNTIL looping instruction. Return from interrupt (RTI) and return from subroutine (RTS) are also provided.

In all family processors other than the ADSP-2100, the IDLE instruction is provided for idling the processor until an interrupt occurs. IDLE puts the processor into a low-power state while waiting for interrupts.

Two addressing modes are supported for memory fetches. Direct addressing uses immediate address values; indirect addressing uses the I registers of the two data address generators (DAGs).

236

Instruction Set Reference 12

The 24-bit instruction word allows a high degree of parallelism in performing operations. The instruction set allows for single-cycle execution of any of the following combinations:

- any ALU, MAC or Shifter operation (conditional or non-conditional)
- any register-to-register move
- any data memory read or write
- a computation with any data register to data register move
- a computation with any memory read or write
- a computation with a read from two memories.

The instruction set allows the programmer maximum flexibility. It provides moves from any register to any other register, and from most registers to/from memory. In addition, almost any ALU, MAC or Shifter operation may be combined with any register-to-register move or with a register move to or from either internal or external memory.

12.3 INSTRUCTION TYPES & NOTATION CONVENTIONS

The ADSP-2100 family instruction set is grouped into the following categories:

- Computational: ALU, MAC, Shifter
- Move
- Program Flow
- Multifunction
- Miscellaneous

Because the multifunction instructions best illustrate the power of the processors' architecture, in the next section we begin with a discussion of this group of instructions.

Throughout this chapter you will find tables summarizing the syntax of the instruction groups. The following notation conventions are used in these tables and in the reference page for each instruction.

Square Brackets [ ]   Anything within square brackets is an optional part of the instruction statement.

Parallel Lines | | Lists of operands are enclosed by vertical parallel bars. One of the operands listed must be chosen. If the parallel bars are within square brackets, then the operand is optional for that instruction.

12 Instruction Set Reference

CAPITAL LETTERS	Capital letters denote a literal in the instruction. Literals are the instruction name (e.g. ADD), register names, or operand selections. Literals must be typed exactly as shown.
operands	Some instruction operands are shown in lowercase letters. These operands may take different values in assembly code. For example, the operand *yop* may be one of several registers: AY0, AY1, or AF.
<exp>	Denotes exponent (shift value) in Shift Immediate instructions; must be an 8-bit signed integer constant.
<data>	Denotes an immediate data value. Can also be a symbol (address label or variable/buffer name) dereferenced by the '%' or '^' operators.
<addr>	Denotes an immediate address value to be encoded in the instruction. The <addr> may be either an immediate value (a constant) or a program label.
<reg>	Refers to any accessible register; see Table 12.7.
<dreg>	Refers to any data register; see Table 12.7.

Immediate values, <exp>, <data>, or <addr>, may be a constant in decimal, hexadecimal, octal or binary format. Default is to decimal.

12.4   MULTIFUNCTION INSTRUCTIONS

Multifunction operations take advantage of the inherent parallelism of the ADSP-2100 family architecture by providing combinations of data moves, memory reads/memory writes, and computation, all in a single cycle.

12.4.1   ALU/MAC With Data & Program Memory Read

Perhaps the single most common operation in DSP algorithms is the sum of products, performed as follows:

- Fetch two operands (such as a coefficient and data point)
- Multiply the operands and sum the result with previous products

238

Instruction Set Reference 12

The ADSP-2100 family processors can execute both data fetches and the multiplication/accumulation in a single-cycle. Typically, a loop of multiply/accumulates can be expressed in ADSP-21xx source code in just two program lines. Since the on-chip program memory (of the ADSP-2101, ADSP-2105, ADSP-2111, and ADSP-21msp50 processors) is fast enough to provide an operand and the next instruction in a single cycle, loops of this type can execute with sustained single-cycle throughput. The on-chip instruction cache of the ADSP-2100 allows the same performance on this processor. An example of such an instruction is:

```
MR=MR+MX0*MY0(SS), MX0=DM(I0,M0), MY0=PM(I4,M5);
```

The first clause of this instruction (up to the first comma) says that MR, the MAC result register, gets the sum of its previous value plus the product of the (current) X and Y input registers of the MAC (MX0 and MY0) both treated as signed (SS).

In the second and third clauses of this multifunction instruction two new operands are fetched. One is fetched from the data memory (DM) pointed to by index register zero (I0, post modified by the value in M0) and the other is fetched from the program memory location (PM) pointed to by I4 (post-modified by M5 in this instance). Note that indirect memory addressing uses a syntax similar to array indexing, with DAG registers providing the index values. Any I register may be paired with any M register within the same DAG.

As discussed in Chapter 2, "Computational Units," registers are read at the beginning of the cycle and written at the end of the cycle. The operands present in the MX0 and MY0 registers at the beginning of the instruction cycle are multiplied and added to the MAC result register, MR. The new operands fetched at the end of this same instruction overwrite the old operands after the multiplication has taken place and are available for computation on the following cycle. You may, of course, load any data registers in conjunction with the computation, not just MAC registers with a MAC operation as in our example.

The computational part of this multifunction instruction may be any unconditional ALU instruction except division or any MAC instruction except saturation. Certain other restrictions apply: the next X operand must be loaded into MX0 from data memory and the new Y operand must be loaded into MY0 from program memory (internal and external memory are identical at the level of the instruction set). The result of the computation must go to the result register (MR or AR) not to the feedback register (MF or AF).

12 Instruction Set Reference

12.4.2　Data & Program Memory Read

This variation of a multifunction instruction is a special case of the multifunction instruction described above in which the computation is omitted. It executes only the dual operand fetch, as shown below:

```
AX0=DM(I2,M0), AY0=PM(I4,M6);
```

In this example we have used the ALU input registers as the destination. As with the previous multifunction instruction, X operands must come from data memory and Y operands from program memory (internal or external memory in either case, for the processors with on-chip memory).

12.4.3　Computation With Memory Read

If a single memory read is performed instead of the dual memory read of the previous two multifunction instructions, a wider range of computations can be executed. The legal computations include all ALU operations except division, all MAC operations and all Shifter operations except SHIFT IMMEDIATE. Computation must be unconditional. An example of this kind of multifunction instruction is:

```
AR=AX0+AY0, AX0=DM(I0,M3);
```

Here an addition is performed in the ALU while a single operand is fetched from data memory. The restrictions are similar to those for previous multifunction instructions. The value of AX0, used as a source for the computation, is the value at the beginning of the cycle. The data read operation loads a new value into AX0 by the end of the cycle. For this same reason, the destination register (AR in the example above) cannot be the destination for the memory read.

12.4.4　Computation With Memory Write

The computation with memory write instruction is similar in structure to the computation with memory read: the order of the clauses in the instruction line, however, is reversed. First the memory write is performed, then the computation, as shown below:

```
DM(I0,M0)=AR, AR=AX0+AY0;
```

Again the value of the source register for the memory write (AR in this example) is the value at the beginning of the instruction. The computation loads a new value into the same register; this is the value in AR at the end of this instruction. Reversing the order of the clauses of the instruction is illegal and causes the assembler to generate a warning; it would imply

Instruction Set Reference 12

that the result of the computation is written to memory when, in fact, the previous value of the register is what is written. There is no requirement that the same register be used in this way although this will usually be the case in order to pipeline operands to the computation.

The restrictions on computation operations are identical to those given above. All ALU operations except division, all MAC operations, and all Shifter operations except SHIFT IMMEDIATE are legal. Computations must be unconditional.

12.4.5   Computation With Data Register Move
This final type of multifunction instruction performs a data register to data register move in parallel with a computation. Most of the restrictions applying to the previous two instructions also apply to this instruction.

```
AR=AX0+AY0, AX0=MR2;
```

Here an ALU addition operation occurs while a new value is loaded into AX0 from MR2. As before, the value of AX0 at the beginning of the instruction is the value used in the computation. The move may be from or to all ALU, MAC and Shifter input and output registers except the feedback registers (AF and MF) and SB.

In the example, the data register move loads the AX0 register with the new value at the end of the cycle. All ALU operations except division, all MAC operations and all Shifter operations except SHIFT IMMEDIATE are legal. Computation must be unconditional.

A complete list of data registers is given in Table 12.7. A complete list of the permissible *xops* and *yops* for computational operations is given in the reference page for each instruction. Table 12.1 shows the legal combinations for multifunction instructions. You may combine operations on the same row with each other.

Unconditional Computations	*Data Move (DM=DAG1)*	*Data Move (PM=DAG2)*
None or any ALU (except Division) or MAC	DM read	PM read
Any ALU except Division	DM read	—
Any MAC	—	PM read
Any Shift except Immediate	DM write	—
	—	PM write
	Register-To-Register	

Table 12.1 Summary Of Valid Combinations For Multifunction Instructions

12 Instruction Set Reference

Multifunction Instructions

```
|<ALU*>†| ,  |AX0|  = DM ( |I0| , |M0|) ,  |AY0|  = PM ( |I4| , |M4|) ;
|<MAC>† |    |AX1|         |I1| , |M1|     |AY1|         |I5| , |M5|
             |MX0|         |I2| , |M2|     |MY0|         |I6| , |M6|
             |MX1|         |I3| , |M3|     |MY1|         |I7| , |M7|


|AX0|  = DM ( |I0| , |M0|) ,  |AY0|  = PM ( |I4| , |M4|) ;
|AX1|         |I1| , |M1|     |AY1|         |I5| , |M5|
|MX0|         |I2| , |M2|     |MY0|         |I6| , |M6|
|MX1|         |I3| , |M3|     |MY1|         |I7| , |M7|


|<ALU>  |  , dreg  =   DM ( |I0| , |M0|) ;
|<MAC>  |                   |I1| , |M1|
|<SHIFT*>|                  |I2| , |M2|
                            |I3| , |M3|
                           ▬▬▬▬▬▬▬▬▬▬▬
                            |I4| , |M4|
                            |I5| , |M5|
                            |I6| , |M6|
                            |I7| , |M7|

                      PM ( |I4| , |M4|)
                           |I5| , |M5|
                           |I6| , |M6|
                           |I7| , |M7|


DM ( |I0| , |M0|)  = dreg,  |<ALU>  |   ;
     |I1| , |M1|            |<MAC>  |
     |I2| , |M2|            |<SHIFT>|
     |I3| , |M3|
    ▬▬▬▬▬▬▬▬▬▬▬
     |I4| , |M4|
     |I5| , |M5|
     |I6| , |M6|
     |I7| , |M7|

PM ( |I4| , |M4|)
     |I5| , |M5|
     |I6| , |M6|
     |I7| , |M7|


|<ALU>  |  , dreg  =   dreg;
|<MAC>  |
|<SHIFT>|
```

Table 12.2 Multifunction Instructions

*All computation must be unconditional; ALU Division and Shift Immediate operations prohibited.

†Result registers, AR and MR, must be used in ALU and MAC operations, not feedback registers (AF and MF).

See section 12.4.1, "ALU/MAC With Data & Program Memory Read."

Instruction Set Reference 12

12.5 ALU, MAC & SHIFTER INSTRUCTIONS

This group of instructions performs computations. All of these instructions can be executed conditionally except the ALU division instructions and the Shifter SHIFT IMMEDIATE instructions.

12.5.1 ALU Group

Here is an example of one ALU instruction, Add/Add with Carry:

```
IF AC AR=AX0+AY0+C;
```

The (optional) conditional expression, IF AC, tests the ALU Carry bit (AC); if there is a carry from the previous instruction, this instruction executes, otherwise a NOP occurs and execution continues with the next instruction. The algebraic expression AR=AX0+AY0+C means that the ALU result register (AR) gets the value of the ALU X input and Y input registers plus the value of the carry-in bit.

Table 12.3 gives a summary list of all ALU instructions. In this list, *condition* stands for all the possible conditions that can be tested and *xop* and *yop* stand for the registers that can be specified as input for the ALU. The conditional clause is optional and is enclosed in square brackets to show this. A complete list of the permissible *xops* and *yops* is given in the reference page for each instruction. A complete list of conditions is given in Table 3.9 of this manual.

ALU Instructions

[IF condition]	AR AF	=	xop	+ yop + C + yop + C	;
[IF condition]	AR AF	=	xop	− yop − yop + C − 1 + C − 1	;
[IF condition]	AR AF	=	yop	− xop − xop + C − 1 − xop + C − 1	;
[IF condition]	AR AF	=	xop	AND OR XOR	yop	;
[IF condition]	AR AF	=	PASS	xop yop −1, 0, 1	;

243

12 Instruction Set Reference

[IF condition]	AR AF	=	–	xop yop	;

[IF condition]	AR AF	=	NOT	xop yop	;

[IF condition]	AR AF	=	ABS	xop	;

[IF condition]	AR AF	=	yop	+ 1	;

[IF condition]	AR AF	=	yop	– 1	;

DIVS yop, xop ;
DIVQ xop ;

Table 12.3 ALU Instructions

12.5.2   MAC Group

Here is an example of one of the MAC instructions, Multiply/Accumulate:

```
IF NOT MV MR=MR+MX0*MY0(UU);
```

The conditional expression, IF NOT MV, tests the MAC overflow bit. If the condition is not true, a NOP is executed. The expression MR=MR+MX0*MY0 is the multiply/accumulate operation: the multiplier result register (MR) gets the value of itself plus the product of the X and Y input registers selected. The modifier in parentheses (UU) treats the operands as unsigned. There can be only one such modifier selected from the available set. (SS) means both are signed, while (US) and (SU) mean that either the first or second operand is signed; (RND) means to round the (implicitly signed) result.

Table 12.4 gives a summary list of all MAC instructions. In this list, *condition* stands for all the possible conditions that can be tested and *xop* and *yop* stand for the registers that can be specified as input for the MAC. A complete list of the permissible *xops* and *yops* is given in the reference page for each instruction.

Instruction Set Reference 12

MAC Instructions

| [IF condition] | MR
MF | = | xop * yop | (| SS
SU
US
UU
RND |); |

| [IF condition] | MR
MF | = | MR + xop * yop | (| SS
SU
US
UU
RND |); |

| [IF condition] | MR
MF | = | MR – xop * yop | (| SS
SU
US
UU
RND |); |

| [IF condition] | MR
MF | = | 0; |

| [IF condition] | MR
MF | = | MR [(RND)]; |

IF MV SAT MR ;

Table 12.4 MAC Instructions

12.5.3   Shifter Group

Here is an example of one of the Shifter instructions, Normalize:

```
IF NOT CE SR= SR OR NORM SI (HI);
```

The conditional expression, IF NOT CE, tests the "not counter expired" condition. If the condition is false, a NOP is executed. The destination of all shifting operations is the Shifter Result register, SR. (The destination of exponent detection instructions is SE or SB, as shown below.) In this example, SI, the Shifter Input register, is the operand. The amount and direction of the shift is controlled by the signed value in the SE register in all shift operations except an immediate shift. Positive values cause left shifts; negative values cause right shifts.

12 Instruction Set Reference

The "SR OR" modifier (which is optional) logically ORs the result with the current contents of the SR register; this allows you to construct a 32-bit value in SR from two 16-bit pieces. "NORM" is the operator and "(HI)" is the modifier that determines whether the shift is relative to the HI or LO (16-bit) half of SR. If "SR OR" is omitted, the result is passed directly into SR.

Table 12.5 gives a summary list of all Shifter instructions. In this list, *condition* stands for all the possible conditions that can be tested.

Shifter Instructions

[IF condition]	SR	=	[SR OR] ASHIFT xop	(HI / LO	);
[IF condition]	SR	=	[SR OR] LSHIFT xop	(HI / LO	);
[IF condition]	SR	=	[SR OR] NORM xop	(HI / LO	);
[IF condition]	SE	=	EXP xop	(HI / LO / HIX	);
[IF condition]	SB	=	EXPADJ xop;		
SR	=		[SR OR] ASHIFT xop BY <exp>	(HI / LO	);
SR	=		[SR OR] LSHIFT xop BY <exp>	(HI / LO	);

Table 12.5 Shifter Instructions

12.6   MOVE: READ & WRITE

MOVE instructions, shown in Table 12.6, move data to and from data registers and external memory. Registers are divided into two groups, referred to as *reg* which includes almost all registers and *dreg*, or data registers, which is a subset. Only the program counter (PC) and the ALU and MAC feedback registers (AF and MF) are not accessible.

Table 12.7 shows which registers belong to these groups. Many of the system control registers are memory-mapped (for the processors with on-chip memory); these registers are read and written as memory locations instead of with register names.

Instruction Set Reference 12

MOVE Instructions

reg   =   reg ;

reg   =   DM (\<address\>) ;

dreg   =   DM (

I0	,	M0	);
I1	,	M1	
I2	,	M2	
I3	,	M3	
I4	,	M4	
I5	,	M5	
I6	,	M6	
I7	,	M7	

DM (

I0	,	M0	)	=	dreg	;
I1	,	M1			\<data\>	
I2	,	M2				
I3	,	M3				
I4	,	M4				
I5	,	M5				
I6	,	M6				
I7	,	M7				

DM (\<address\>) =   reg;

reg   =   \<data\> ;

dreg   =   PM (

I4	,	M4	);
I5	,	M5	
I6	,	M6	
I7	,	M7	

PM (

I4	,	M4	)	=	dreg;
I5	,	M5			
I6	,	M6			
I7	,	M7			

Table 12.6 MOVE Instructions

12 Instruction Set Reference

Registers: **reg**

SB	*Data Registers:* **dreg**
PX	
I0 – I7, M0 – M7, L0 – L7	AX0, AX1, AY0, AY1, AR
CNTR	MX0, MX1, MY0, MY1, MR0, MR1, MR2
ASTAT, MSTAT, SSTAT	SI, SE, SR0, SR1
IMASK, ICNTL, IFC	
TX0, TX1, RX0, RX1	

Table 12.7 Processor Registers: reg & dreg

12.7 PROGRAM FLOW CONTROL

Program flow control on the ADSP-2100 family processors is simple but powerful. Here is an example of one instruction:

```
IF EQ JUMP my_label;
```

JUMP, of course, is a familiar construct from many other processors. *My_label* is any identifier you wish to use as a label for the destination jumped to. Instead of the label, an index register in DAG2 may be explicitly used. The default scope for any label is the source code module in which it is declared. The assembler directive .ENTRY makes a label visible as an entry point for routines outside the module. Conversely, the .EXTERNAL directive makes it possible to use a label declared in another module.

If the counter condition (CE, NOT CE) is to be used, an assignment to CNTR must be executed to initialize the counter value. JUMP and CALL permit the additional conditionals "FLAG_IN" and "NOT FLAG_IN" to be used for branching on the state of the FI pin, but only with direct addressing, not with DAG2 as the address source.

RTS (return from subroutine) and RTI (return from interrupt) provide for conditional return from CALL or interrupt vectors respectively.

The IDLE instruction provides a way to wait for interrupts. IDLE causes the processor to wait in a low-power state until an interrupt occurs. When an interrupt is serviced, control returns to the instruction following the IDLE statement. IDLE uses less power than loops created with JUMP.

Table 12.8 gives a summary of all program flow control instructions. The *condition* and *termination* terms are described in Chapter 4, Tables 3.9 and 3.1. (Note that neither the "FLAG_IN" and "NOT FLAG_IN" conditions

248

for JUMP and CALL nor the IDLE instruction are implemented on the ADSP-2100 processor.)

Program Flow Control Instructions

[IF condition]	JUMP	(I4)	;
		(I5)	
		(I6)	
		(I7)	
		<address>	

| IF | FLAG_IN | JUMP | <address> ; |
| | NOT FLAG_IN | | |

[IF condition]	CALL	(I4)	;
		(I5)	
		(I6)	
		(I7)	
		<address>	

| IF | FLAG_IN | CALL | <address> ; |
| | NOT FLAG_IN | | |

[IF condition]   RTS ;

[IF condition]   RTI ;

DO <address> [UNTIL termination] ;

IDLE;

Table 12.8 Program Flow Control Instructions

12.8   MISCELLANEOUS INSTRUCTIONS

There are several miscellaneous instructions. NOP is a no operation instruction. The PUSH/POP instructions allows you to explicitly control the status, counter, PC and loop stacks; interrupt servicing automatically pushes and pops some of these stacks.

The Mode Control instruction enables and disables processor modes of operation. The instruction governs modes common to all family processors (bit-reversal on DAG1, latching ALU overflow, saturating the ALU result register, choosing the primary or secondary register set), as well as the modes implemented on the ADSP-2101, ADSP-2105, ADSP-2111, and ADSP-21msp50 (GO mode for continued operation during bus grant, multiplier shift mode for fractional or integer arithmetic, and timer enabling).

12 Instruction Set Reference

A single ENA or DIS can be followed by any number of mode identifiers, separated by commas; ENA and DIS can also be repeated. All seven modes can be enabled, disabled, or changed in a single instruction.

The MODIFY instruction modifies the address pointer in the I register selected with the value in the selected M register, without performing any actual memory access. As always, the I and M registers must be from the same DAG; any of I0-I3 may be used only with one from M0-M3 and the same for I4-I7 and M4-M7. If circular buffering is in use, modulus logic applies (See Chapter 4, "Data Transfer," for more information).

The FO (Flag Out), FL0, FL1 and FL2 pins can each be set, cleared, or toggled. This instruction provides a control structure for multiprocessor communication.

Miscellaneous Instructions

NOP;

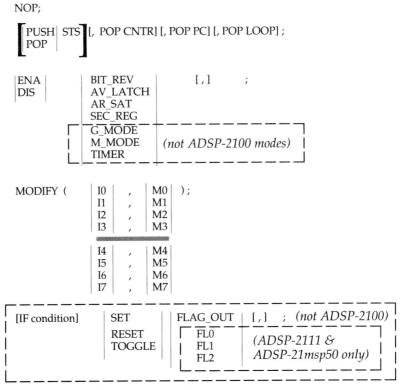

Table 12.9 Miscellaneous Instructions

Instruction Set Reference 12

12.9   DATA STRUCTURES & VARIABLES

The ADSP-2100 Family Development Software supports the declaration and use of a simple data structure: one-dimensional arrays. The array may contain a single value (a variable) or multiple values (an array). In addition, the array may be used as a circular buffer. Here is a brief discussion of each instance, with an example of how they are declared and used. Complete syntax for these and other assembler directives is given in the *ADSP-2100 Family Assembler Manual*.

12.9.1   Arrays

Arrays are the basic data structure of the ADSP-2100 family. In our literature, the word "array" and the expression "data buffer" (as well as "variable") are used interchangeably. Arrays are declared with assembler directives and can be referenced indirectly and by name, can be initialized from immediate values in a directive or from external data files, and can be linear or circular with automatic wraparound.

An array is declared with a directive such as

```
.VAR/DM coefficients[128];
```

This declares an array of 128 16-bit values located in data memory (DM). The special operators ^ and % reference the address and length, respectively, of the array. It could be referenced as shown below:

```
I0 = ^coefficients {point to address of buffer}
MX0=DM(I0,M0);     {load MX0 from buffer}
```

These instructions load a value into MX0 from the beginning of the *coefficients* buffer in data memory. With the automatic post-modify of the DAGs, you could execute the second of these instructions in a loop and continuously advance through the buffer.

Alternatively, when you only need to address the first location, you can directly use the buffer name as a label in many circumstances such as

```
MX0=DM(coefficients);
```

12 Instruction Set Reference

The linker substitutes the actual address for the label. It is also possible to initialize a complete array/buffer from a data file, using the .INIT directive:

```
.INIT coefficients: <filename.dat>;
```

This assembler directive reads the values from the file *filename.dat* into the array at link time. This feature is supported only in the simulator — data cannot be loaded directly into on-chip data memory by the hardware booting sequence.

An array or data buffer with a length of one is a simple single-word variable, and is declared in this way:

```
.VAR/DM coefficient;
```

12.9.2   Circular Arrays/Buffers

A common requirement in DSP is the circular buffer. This is directly implemented by the processors' data address generators (DAGs), using the L (length) registers. First, you must declare the buffer as circular:

```
.VAR/DM/CIRC coefficients[128];
```

This identifies it to the linker for placement on the proper address boundary. Next, you must initialize the L register, typically using the assemblers's % operator (or a constant) and, in the example below, the I register and M register:

```
L0= %coefficients;    {length of circular buffer}
I0= ^coefficients;    {point to address of buffer}
M0= 1;                {increment by 1 location each time}
```

Now a statement like

```
MX0=DM(I0,M0);        {load MX0 from buffer}
```

placed in a loop, cycles continuously through *coefficients* and wraps around automatically. L registers should be initialized to zero for buffers of any length that are not circular.

Instruction Set Reference 12

12.10   INDIRECT LINEAR (NON-CIRCULAR) ADDRESSING

The ADSP-2100 family processors allow two addressing modes for data memory fetches: direct and register indirect. Indirect addressing is accomplished by loading an address into an I (index) register and specifying one of the available M (modify) registers.

The L registers are provided to facilitate wraparound addressing of circular data buffers. A circular buffer is only implemented when an L register is set to a non-zero value. For linear (i.e. non-circular) indirect addressing, the L register corresponding to the I register used **must be set to zero**. Do not assume that the L registers are automatically initialized or may be ignored; the I, M, and L registers contain random values following processor reset. Your program must initialize the L registers corresponding to any I registers it uses.

For further details on this topic, refer to Chapter 4, "Data Transfer."

12.10.1 Set L Registers To 0 For Non-Circular Addressing

Setting an L register to a non-zero value activates the processor's circular addressing modulus logic. For linear indirect addressing you must set the appropriate L register to zero to disable the modulus logic.

Here is a simple example of linear indirect addressing:

```
I3=0x3800;
M3=0;
L3=0;
AX0=DM(I3,M3);
```

Here is an example which uses a memory variable to store an address pointer:

```
.VAR/DM/RAM  addr_ptr;  {variable holds address to be accessed}
I3=DM(addr_ptr);        {I3 loaded using direct addressing}
L3=0;                   {disable circular addressing}
M1=0;                   {no post-modify of I3}
AX0=DM(I3,M1);          {AX0 loaded using indirect addressing}
```

12 Instruction Set Reference

12.11   CYCLE TIME NOTES

All instructions execute in a single cycle except under certain conditions, as explained below.

12.11.1 ADSP-2100 Extra Cycle Conditons

If an instruction causes a data fetch from program memory, an extra cycle is required since the processor cannot pre-fetch the next instruction in the same cycle. If, however, the next instruction is available from the cache, the extra cycle is not needed.

In many cases, such as inside loops, the overhead cycle is incurred only once as the instruction can subsequently be fetched from the cache.

12.11.2   Extra Cycle Conditions For All Other Processors

The ADSP-2101, ADSP-2105, ADSP-2111, and ADSP-21msp50 generate extra cycles under different conditions than the ADSP-2100.

12.11.2.1 Multiple Off-Chip Memory Accesses

While the ADSP-2100 has separate off-chip data and address busses for program and data memory, the data and address busses of all other ADSP-2100 family processors are multiplexed off-chip.

Because of this, these processors can perform only one off-chip access per instruction in a single cycle. If two off-chip accesses are required—the instruction fetch and one data fetch, for example, or data fetches from both program and data memory—then one overhead cycle occurs. If three off-chip accesses are required—the instruction fetch as well as data fetches from both program and data memory—then two overhead cycles occur.

A multifunction instruction requires three items to be fetched from memory: the instruction itself and two data words. No extra cycle is needed to execute the instruction as long as only one of the fetches is from external memory. (Two fetches must be from on-chip memory, either PM or DM.)

12.11.2.2 Wait States

All family processors except the ADSP-2100 allow the programming of wait states for external memory chips. (The DMACK signal of the ADSP-2100 is used for this purpose.) Up to seven extra wait state cycles may be added to the processor's access time for external memory. Extra cycles inserted due to wait states are in addition to any caused by multiple off-chip accesses (as described above). Wait state programming is described in Chapter 10, "Memory Interface."

Instruction Set Reference   12

Wait states and multiple off-chip memory accesses are the two cases when an extra cycle is generated during instruction execution. The following case, SPORT autobuffering, causes the insertion of extra cycles between instructions.

12.11.2.3 SPORT Autobuffering
If serial port autobuffering is being used to transfer data words to or from data memory, then one memory access is "stolen" for each transfer. The stolen memory access occurs only between complete instructions. If extra cycles are required to execute any instruction (for one of the two reasons above), the processor waits until it is completed before "stealing" the access cycle. Autobuffering is described in the Serial Ports chapter of this manual.

12.12   INSTRUCTION SET SYNTAX
The following sections describe instruction set syntax and other notation conventions used in the reference page of each instruction.

12.12.1 Punctuation & Multifunction Instructions
All instructions terminate with a semicolon. A comma separates the clauses of a multifunction instruction but does not terminate it. For example, the statements below in Example A comprise one multifunction instruction (which can execute in a single cycle). Example B shows two separate instructions, requiring two instruction cycles.

Example A: One multifunction instruction

```
AX0 = DM(I0, M0),  a comma is used in multifunction instructions
AY0 = PM(I4, M4);
```

Example B: Two separate instructions

```
AX0 = DM(I0, M0);  a semicolon terminates an instruction
AY0 = PM(I4, M4);
```

12 Instruction Set Reference

12.12.2 Syntax Notation Example

Here is an example of one instruction, the ALU Add/Add with Carry instruction:

$$[\text{ IF cond }] \quad \begin{vmatrix} \text{AR} \\ \text{AF} \end{vmatrix} = \text{xop} + \begin{vmatrix} \text{yop} \\ \text{C} \\ \text{yop} + \text{C} \end{vmatrix} \quad ;$$

The permissible *conds*, *xops* and *yops* are given in a list. The conditional IF clause is enclosed in square brackets, indicating that it is optional.

The destination register for the add operation must be either AR or AF. These are listed within parallel bars, indicating that one of the two must be chosen.

Similarly, the *yop* term may consist of a Y operand, the carry bit, or the sum of both. One of these three terms must be used.

12.12.3 Status Register Notation

The following notation is used in the discussion of the effect each instruction has on the processors' status registers:

* An asterisk indicates a bit in the status word that is changed by the execution of the instruction.

– A dash indicates that a bit is not affected by the instruction.

0 or 1 Indicates that a bit is unconditionally cleared or set.

For example, the status word ASTAT is shown below:

ASTAT:	7	6	5	4	3	2	1	0
	SS	MV	AQ	AS	AC	AV	AN	AZ
	–	*	–	–	–	0	–	–

Here the MV bit is updated and the AV bit is cleared.

256

Instruction Set Reference   12

12.13   PROGRAM EXAMPLE

This section presents an example of an FIR filter program written for the
ADSP-2111 with discussion of each part of the program. The program can
also be executed on the ADSP-2101, ADSP-2105, or ADSP-21msp50, with
minor modifications. This FIR filter program demonstrates much of the
conceptual power of the ADSP-2100 family architecture and instruction
set. More complex programs would, of course, use additional features of
the language.

```
     {ADSP-2111 FIR Filter Routine

        -serial port 0 used for I/O
        -internally generated serial clock
        -12.288 MHz processor clock rate is divided to 1.536 MHz serial clock
        -serial clock divided to 8 kHz frame sampling rate}

     .MODULE/RAM/ABS=0          main_routine;     {program loaded from EPROM, MMAP=0}
A    .INCLUDE                   <const.h>;
     .VAR/DM/RAM/ABS=0x3800/CIRC data_buffer[taps];  {on-chip data buffer}
B    .VAR/PM/RAM/CIRC           coefficient[taps];
     .GLOBAL                    data_buffer, coefficient;
     .EXTERNAL                  fir_start;
     .INIT                      coefficient:<coeff.dat>;

     {code starts here}
     {load interrupt vector addresses}

C              JUMP restarter; NOP; NOP; NOP;    {restart interrupt}
               RTI; NOP; NOP; NOP;               {IRQ2 interrupt}
               RTI; NOP; NOP; NOP;               {HIP write interrupt}
               RTI; NOP; NOP; NOP;               {HIP read interrupt}
               RTI; NOP; NOP; NOP;               {SPORT0 transmit int}
               JUMP fir_start; NOP; NOP; NOP;    {SPORT0 receive int}
               RTI; NOP; NOP; NOP;               {SPORT1 transmit int}
               RTI; NOP; NOP; NOP;               {SPORT1 receive int}
               RTI; NOP; NOP; NOP;               {TIMER interrupt}

D    {initializations}

     restarter:  L0=%data_buffer;        {setup circular buffer length}
                 L4=%coefficient;        {setup circular buffer length}

                 M0=1;                   {modify=1 for increment through buffers}
                 M4=1;

                 I0=^data_buffer;        {point to data start}
                 I4=^coefficient;        {point to coeff start}

                 CNTR=%data_buffer;
                 DO clear UNTIL CE;       {clear data buffer}
     clear:      DM(I0,M0)=0;
```

257

12 Instruction Set Reference

```
E  {set up memory-mapped control registers}

               AX0=191;
               DM(0x3FF4)=AX0;          {set up divide value for 8KHz RFS}
               AX0=3;
               DM(0x3FF5)=AX0;          {1.536MHz internal serial clock}
               AX0=0x69B7;
               DM(0x3FF6)=AX0;          {multichannel disabled}
                                        {internally generated serial clock}
                                        {receive frame sync required}
                                        {receive width 0}
                                        {transmit frame sync required}
                                        {transmit width 0}
                                        {int transmit frame sync disabled}
                                        {int receive frame sync enabled}
                                        {u-law companding}
                                        {8 bit words}

               AX0=0x7000;
               DM(0x3FFE)=AX0;          {DM wait states: }
                                        {  0x3400-0x37FF 7 waits}
                                        {  all else 0 waits}

               AX0=0x1000;
               DM(0x3FFF)=AX0;          {SPORT0 enabled}
                                        {boot from boot page 0}
                                        {0 PM waits}
                                        {0 boot memory waits}

               ICNTL = 0x00;
               IMASK = 0x0018;          {enable SPORT0 interrupt only}
mainloop:      IDLE;                    {wait for interrupt}
               JUMP mainloop;

.ENDMOD;
```

Setup and Main Loop Routine

```
.CONST              taps = 15, taps_less_one = 14;
```

Include File, Constant Initialization

Listing 12.1 Program Example Listing, Main Routine & Constants File

12.13.1 Example Program: Setup Routine Discussion

The setup and main loop routine, shown in Listing 12.1, performs initialization and then loops on the IDLE instruction to wait until the receive interrupt from SPORT0 occurs. The filter is interrupt-driven. When the interrupt occurs control shifts to the interrupt service routine (shown in Listing 12.2).

Line A of the program shows that the constant declarations are contained in a separate file.

Instruction Set Reference 12

Section B of the program includes the assembler directives defining two circular buffers in on-chip memory: one in data memory RAM (used to hold a delay line of samples) and one in program memory RAM (used to store coefficients for the filter). The coefficients are actually loaded from an external file by the linker. These values can be changed without reassembling; only another linking is required.

Section C shows the setup of interrupts. Since this code module is located at absolute address zero (as indicated by the ABS qualifier in the .MODULE directive), the first instruction is placed at the restart vector: address 0x0000. The first location is the restart vector instruction, which jumps to the routine *restarter*. Interrupt vectors that are not used are filled with a return from interrupt instruction followed by NOPs. (Since only one interrupt will be enabled, this is only a thorough programming practice rather than a necessity.) The SPORT0 receive interrupt vector jumps to the interrupt service routine.

Section D, *restarter*, sets up the index (I), length (L), and modify (M) registers used to address the two circular buffers. A non-zero value for length activates the processor's modulus logic. Each time the interrupt occurs, the I register pointers advance one position through the buffers. The *clear* loop zeroes all values in the data memory buffer.

Section E, after *clear*, sets up the processor's memory-mapped control registers used in this system. See Appendix E for a summary listing of control register initialization.

SPORT0 is set up to generate the serial clock internally at 1.536 MHz, based on a processor clock rate of 12.288 MHz. The RFS and TFS signals are both required and the RFS signal is generated internally at 8 kHz, while the TFS signal comes from the external device communicating with the processor.

Finally, SPORT0 is enabled and the interrupts are enabled. Now the IDLE instruction causes the processor to wait for interrupts. After the return from interrupt instruction, execution resumes at the instruction following the IDLE instruction. Once these setup instructions have been executed, all further activity takes place in the interrupt service routine, shown in Listing 12.2.

12   Instruction Set Reference

```
.MODULE/ROM fir_routine;           {relocatable FIR interrupt module}
.INCLUDE  <const.h>;               {include constant declarations}
.ENTRY    fir_start;               {make label visible outside module}
.EXTERNAL data_buffer, coefficient; {make globals accessible in module}

{interrupt service routine code}

FIR_START:    CNTR = taps_less_one;                  {N-1 passes within DO UNTIL}
              SI = RX0;                               {read from SPORT0}
              DM(I0,M0) = SI;                         {transfer data to buffer}
              MR=0, MY0=PM(I4,M4), MX0=DM(I0,M0);     {set up multiplier for loop}
              DO convolution UNTIL CE;                {CE = counter expired}
convolution:  MR=MR+MX0*MY0(SS), MY0=PM(I4,M4), MX0=DM(I0,M0);
                                                      {MAC these, fetch next}
              MR=MR+MX0*MY0(RND);                     {Nth pass with rounding}
              TX0 = MR1;                              {write to sport}
              RTI;                                    {return from interrupt}
.ENDMOD;
```

Listing 12.2 Program Example Listing, Interrupt Routine

12.13.2   Example Program: Interrupt Routine Discussion

This subroutine transfers the received data to the next location in the circular buffer (overwriting the oldest sample). All samples and coefficients are then multiplied and the products are accumulated to produce the next output value. The subroutine checks for overflow and saturates the output value to the appropriate full scale, then writes the result to the transmit section of SPORT0 and returns.

The first four lines of the listing declare the code module (which is relocatable rather than placed at an absolute address), include the same file of constants, and make the entry point visible to the main routine with the .ENTRY directive. Likewise, the .EXTERNAL directive makes the main routine labels visible in the interrupt routine.

The subroutine begins by loading the counter register (CNTR). The new sample is read from SPORT0's receive data register, RX0, into the SI register; the choice of SI is of no particular significance. Then, the data is written into the data buffer. Because of the automatic circular buffer addressing, the new data overwrites the oldest sample. The N-most recent samples are always in the buffer.

Instruction Set Reference 12

The fourth instruction of the routine, `MR=0, MY0=PM(I4,M4),` `MX0=DM(I0,M0)`, zeroes the multiplier result register (MR) and fetches the first two operands. This instruction accesses both program and data memory but still executes in a single cycle because of the processor's architecture.

The *convolution* label identifies the loop itself, consisting of only two instructions, one setting up the loop (DO UNTIL) and one "inside" the loop. The MAC instruction multiplies and accumulates the previous set of operands while fetching the next ones from each memory. This instruction also accesses both memories.

The final value is transferred back to SPORT0, to the transmit data register TX0, to be sent to the communicating device.

12 Instruction Set Reference

Syntax: [IF cond] $\left|\begin{array}{c} AR \\ AF \end{array}\right|$ = xop $\left|\begin{array}{l} + yop \\ + C \\ + yop + C \end{array}\right|$;

Permissible xops		Permissible yops	Permissible conds		
AX0	MR2	AY0	EQ	LE	AC
AX1	MR1	AY1	NE	NEG	NOT AC
AR	MR0	AF	GT	POS	MV
	SR1		GE	AV	NOT MV
	SR0		LT	NOT AV	NOT CE

Example: IF EQ AR = AX0 + AY0 + C;

Description: Test the optional condition and, if true, perform the specified addition. If false then perform a no-operation. Omitting the condition performs the addition unconditionally. The addition operation adds the first source operand to the second source operand along with the ALU carry bit, AC, (if designated by the "+C" notation), using binary addition. The result is stored in the destination location. The operands are contained in the data registers specified in the instruction.

Status Generated:

ASTAT:	7	6	5	4	3	2	1	0
	SS	MV	AQ	AS	AC	AV	AN	AZ
	–	–	–	–	*	*	*	*

AZ   Set if the result equals zero. Cleared otherwise.
AN   Set if the result is negative. Cleared otherwise.
AV   Set if an arithmetic overflow occurs. Cleared otherwise.
AC   Set if a carry is generated. Cleared otherwise.

Instruction Format:

Conditional ALU/MAC operation, Instruction Type 9:

23 22 21 20 19	18	17 16 15 14 13	12 11	10 9 8	7 6 5 4	3 2 1 0
0 0 1 0 0	Z	AMF	Yop	Xop	0 0 0 0	COND

AMF specifies the ALU or MAC operation, in this case:
    AMF = 10010 for xop + yop + C operation
    AMF = 10011 for xop + yop
Note that xop + C is a special case of xop + yop + C with yop = 0

Z: Destination register   Yop: Y operand
Xop: X operand   COND: condition

ALU
SUBTRACT X-Y / SUBTRACT X-Y with BORROW

Syntax: [IF cond] $\left| \begin{array}{c} AR \\ AF \end{array} \right|$ = xop $\left| \begin{array}{l} -\text{yop} \\ -\text{yop} + C{-}1 \\ +\,C{-}1 \end{array} \right|$;

Permissible xops		*Permissible yops*	*Permissible conds*		
AX0	MR2	AY0	EQ	LE	AC
AX1	MR1	AY1	NE	NEG	NOT AC
AR	MR0	AF	GT	POS	MV
	SR1		GE	AV	NOT MV
	SR0		LT	NOT AV	NOT CE

Example: IF GE AR = AX0 – AY0;

Description: Test the optional condition and, if true, then perform the specified subtraction. If the condition is not true then perform a no-operation. Omitting the condition performs the subtraction unconditionally. The subtraction operation subtracts the second source operand from the first source operand, and optionally adds the ALU Carry bit (AC) minus 1 (H#0001), and stores the result in the destination location. The (C-1) quantity effectively implements a borrow capability for multiprecision subtractions. The operands are contained in the data registers specified in the instruction.

Status Generated:

ASTAT:	7	6	5	4	3	2	1	0
	SS	MV	AQ	AS	AC	AV	AN	AZ
	-	-	-	-	*	*	*	*

AZ	Set if the result equals zero. Cleared otherwise.
AN	Set if the result is negative. Cleared otherwise.
AV	Set if an arithmetic overflow occurs. Cleared otherwise.
AC	Set if a carry is generated. Cleared otherwise.

Instruction Format:

Conditional ALU/MAC operation, Instruction type 9:

23 22 21 20 19 18	17 16 15 14 13	12 11	10 9 8	7 6 5 4	3 2 1 0
0 0 1 0 0 Z	AMF	Yop	Xop	0 0 0 0	COND

AMF specifies the ALU or MAC operation. In this case,
    AMF = 10110 for xop - yop + C - 1 operation.
    AMF = 10111 for xop - yop operation.
Note that xop + C - 1 is a special case of xop - yop + C - 1 with yop=0.

Z:	Destination register	Yop:	Y operand
Xop:	X operand	COND:	condition

Syntax:

$$[\text{ IF cond }] \left| \begin{array}{c} AR \\ AF \end{array} \right| = \left| yop - \left| \begin{array}{c} xop \\ xop + C - 1 \\ -xop + C - 1 \end{array} \right| \right| ;$$

Permissible xops		*Permissible yops*	*Permissible conds*		
AX0	MR2	AY0	EQ	LE	AC
AX1	MR1	AY1	NE	NEG	NOT AC
AR	MR0	AF	GT	POS	MV
	SR1		GE	AV	NOT MV
	SR0		LT	NOT AV	NOT CE

Example: IF GT AR = AY0 – AX0 + C – 1;

Description: Test the optional condition and, if true, then perform the specified subtraction. If the condition is not true then perform a no-operation. Omitting the condition performs the subtraction unconditionally. The subtraction operation subtracts the second source operand from the first source operand, optionally adds the ALU Carry bit (AC) minus 1 (H#0001), and stores the result in the destination location. The (C-1) quantity effectively implements a borrow capability for multiprecision subtractions. The operands are contained in the data registers specified in the instruction.

Status Generated:

ASTAT:	7	6	5	4	3	2	1	0
	SS	MV	AQ	AS	AC	AV	AN	AZ
	–	–	–	–	*	*	*	*

AZ   Set if the result equals zero. Cleared otherwise.
AN   Set if the result is negative. Cleared otherwise.
AV   Set if an arithmetic overflow occurs. Cleared otherwise.
AC   Set if a carry is generated. Cleared otherwise.

Instruction Format:

Conditional ALU/MAC Operation, Instruction Type 9:

23	22	21	20	19	18	17 16 15 14 13	12 11	10 9 8	7 6 5 4	3 2 1 0
0	0	1	0	0	Z	AMF	Yop	Xop	0 0 0 0	COND

AMF specifies the ALU or MAC operation. In this case,
    AMF = 11010 for yop - xop + C - 1 operation.
    AMF = 11001 for yop - xop operation.
Note that –xop + C - 1 is a special case of yop - xop + C - 1 with yop=0.

Z:	Destination register	Yop:	Y operand
Xop:	X operand	COND:	condition

12 ALU
AND, OR, XOR

Syntax: [IF cond] $\begin{vmatrix} AR \\ AF \end{vmatrix}$ = xop $\begin{vmatrix} AND \\ OR \\ XOR \end{vmatrix}$ yop ;

Permissible xops		*Permissible yops*	*Permissible conds*		
AX0	MR2	AY0	EQ	LE	AC
AX1	MR1	AY1	NE	NEG	NOT AC
AR	MR0	AF	GT	POS	MV
	SR1		GE	AV	NOT MV
	SR0		LT	NOT AV	NOT CE

Example: AR = AX0 XOR AY0;

Description: Test the optional condition and if true, then perform the specified bitwise logical operation (logical AND, Inclusive OR, or EXCLUSIVE OR). If the condition is not true then perform a no-operation. Omitting the condition performs the logical operation unconditionally. The operands are contained in the data registers specified in the instruction.

Status Generated:

ASTAT:	7	6	5	4	3	2	1	0
	SS	MV	AQ	AS	AC	AV	AN	AZ
	–	–	–	–	0	0	*	*

AZ	Set if the result equals zero. Cleared otherwise.
AN	Set if the result is negative. Cleared otherwise.
AV	Always cleared.
AC	Always cleared.

Instruction Format:

Conditional ALU/MAC Operation, Instruction Type 9:

23 22 21 20 19	18	17 16 15 14 13	12 11	10 9 8	7 6 5 4	3 2 1 0
0 0 1 0 0	Z	AMF	Yop	Xop	0 0 0 0	COND

AMF specifies the ALU or MAC operation. In this case,
   AMF = 11100 for AND operation.
   AMF = 11101 for OR operation.
   AMF = 11110 for XOR operation.

Z:	Destination register	Yop:	Y operand
Xop:	X operand	COND:	condition

Syntax: [IF cond] $\left| \begin{array}{c} AR \\ AF \end{array} \right|$ = PASS $\left| \begin{array}{c} xop \\ yop \end{array} \right|$;

Permissible xops		*Permissible yops*	*Permissible conds*		
AX0	MR2	AY0	EQ	LE	AC
AX1	MR1	AY1	NE	NEG	NOT AC
AR	MR0	AF	GT	POS	MV
	SR1	−1, 0, 1	GE	AV	NOT MV
	SR0		LT	NOT AV	NOT CE

Example: IF GE AR = PASS AY0;

Description: Test the optional condition and if true, pass the source operand unmodified through the ALU block and store in the destination location. If the condition is not true perform a no-operation. Omitting the condition performs the PASS unconditionally. The source operand is contained in the data registers specified in the instruction.

The PASS instruction performs the transfer to the AR register and affects the status flag; this instruction is different from a register move operation which does not affect any status flags. PASS 0 is one method of clearing AR. PASS 0 can also be combined in a multifunction instruction in conjunction with memory reads and writes to clear AR.

Status Generated:

ASTAT:	7	6	5	4	3	2	1	0
	SS	MV	AQ	AS	AC	AV	AN	AZ
	−	−	−	−	0	0	*	*

AZ   Set if the result equals zero. Cleared otherwise.
AN   Set if the result is negative. Cleared otherwise.
AV   Always cleared.
AC   Always cleared.

Instruction Format:
Conditional ALU/MAC Operation, Instruction Type 9:

23 22 21 20 19 18	17 16 15 14 13	12 11	10 9 8	7 6 5 4	3 2 1 0
0 0 1 0 0 Z	AMF	Yop	Xop	0 0 0 0	COND

AMF specifies the ALU or MAC operation. In this case,
    AMF = 10000 for PASS yop operation.
    AMF = 10011 for PASS xop operation.
    AMF = 10001 for PASS 1 operation.
    AMF = 11000 for PASS −1 operation.
Note that PASS xop is a special case of xop + yop, with yop=0.
Note that PASS 1 is a special case of yop + 1, with yop=0.
Note that PASS −1 is a special case of yop - 1, with yop=0.

Z:	Destination register	Yop:	Y operand
Xop:	X operand	COND:	condition

12 ALU NEGATE

Syntax: [IF cond] | $\begin{matrix} AR \\ AF \end{matrix}$ | $= -$ | $\begin{matrix} xop \\ yop \end{matrix}$ | ;

Permissible xops		*Permissible yops*	*Permissible conds*		
AX0	MR2	AY0	EQ	LE	AC
AX1	MR1	AY1	NE	NEG	NOT AC
AR	MR0	AF	GT	POS	MV
	SR1		GE	AV	NOT MV
	SR0		LT	NOT AV	NOT CE

Example: IF LT AR = – AY0;

Description: Test the optional condition and if true, then NEGATE the source operand and store in the destination location. If the condition is not true then perform a no-operation. Omitting the condition performs the NEGATE operation unconditionally. The source operand is contained in the data register specified in the instruction.

Status Generated:

ASTAT:	7	6	5	4	3	2	1	0
	SS	MV	AQ	AS	AC	AV	AN	AZ
	–	–	–	–	*	*	*	*

AZ	Set if the result equals zero. Cleared otherwise.
AN	Set if the result is negative. Cleared otherwise.
AV	Set if operand = H#8000. Cleared otherwise.
AC	Set if operand equals zero. Cleared otherwise.

Instruction Format:

Conditional ALU/MAC Operation, Instruction Type 9:

23 22 21 20 19	18	17 16 15 14 13	12 11	10 9 8	7 6 5 4	3 2 1 0
0 0 1 0 0	Z	AMF	Yop	Xop	0 0 0 0	COND

AMF specifies the ALU or MAC operation. In this case,
    AMF = 10101 for - yop operation.
    AMF = 11001 for - xop operation
Note that –xop is a special case of yop –xop, with yop specified to be 0.

Z:	Destination register	Yop:	Y operand
Xop:	X operand	COND:	condition

268

Syntax: [IF cond] $\left|\begin{array}{c} \text{AR} \\ \text{AF} \end{array}\right|$ = NOT $\left|\begin{array}{c} \text{xop} \\ \text{yop} \end{array}\right|$;

Permissible xops		Permissible yops	Permissible conds		
AX0	MR2	AY0	EQ	LE	AC
AX1	MR1	AY1	NE	NEG	NOT AC
AR	MR0	AF	GT	POS	MV
	SR1	0	GE	AV	NOT MV
	SR0		LT	NOT AV	NOT CE

Example: IF NE AF = NOT AX0;

Description: Test the optional condition and if true, then perform the logical complement (ones complement) of the source operand and store in the destination location. If the condition is not true then perform a no-operation. Omitting the condition performs the complement operation unconditionally. The source operand is contained in the data register specified in the instruction.

Status Generated:

ASTAT:	7	6	5	4	3	2	1	0
	SS	MV	AQ	AS	AC	AV	AN	AZ
	–	–	–	–	0	0	*	*

AZ   Set if the result equals zero. Cleared otherwise.
AN   Set if the result is negative. Cleared otherwise.
AV   Always cleared.
AC   Always cleared.

Instruction Format:
Conditional ALU/MAC Operation, Instruction Type 9:

23	22	21	20	19	18	17	16	15	14	13	12	11	10	9	8	7	6	5	4	3	2	1	0
0	0	1	0	0	Z			AMF				Yop			Xop			0	0	0	0	COND	

AMF specifies the ALU or MAC operation. In this case,
    AMF = 10100 for NOT yop operation.
    AMF = 11011 for NOT xop operation.

Z: Destination register   Yop: Y operand
Xop: X operand   COND: condition

12 ALU
ABSOLUTE VALUE

Syntax: [IF cond] $\begin{vmatrix} AR \\ AF \end{vmatrix}$ = ABS xop ;

Permissible xops   *Permissible conds*
AX0   MR2   EQ   LE   AC
AX1   MR1   NE   NEG   NOT AC
AR   MR0   GT   POS   MV
      SR1   GE   AV   NOT MV
      SR0   LT   NOT AV   NOT CE

Example: IF NEG AF = ABS AX0 ;

Description: Test the optional condition and, if true, then take the absolute value of the source operand and store in the destination location. If the condition is not true then perform a no-operation. Omitting the condition performs the absolute value operation unconditionally. The source operand is contained in the data register specified in the instruction.

Status Generated:

ASTAT: 7   6   5   4   3   2   1   0
             SS   MV   AQ   AS   AC   AV   AN   AZ
             -    -   -   *   0   *   *   *

AZ   Set if the result equals zero. Cleared otherwise.
AN   Set if *xop* is H#8000. Cleared otherwise.
AV   Set if *xop* is H#8000. Cleared otherwise.
AC   Always cleared.
AS   Set if the source operand is negative. Cleared otherwise.

Instruction Format:

Conditional ALU/MAC Operation, Instruction Type 9:

23 22 21 20 19	18	17 16 15 14 13 12 11	10 9	8 7 6 5	4 3 2 1 0
0 0 1 0 0	Z	AMF	0 0	Xop	0 0 0 0	COND

AMF specifies the ALU or MAC operation. In this case,
    AMF = 11111 for ABS xop operation.

Z: Destination register
Xop: X operand   COND: condition

Syntax: [IF cond] $\left| \begin{matrix} AR \\ AF \end{matrix} \right|$ = yop + 1 ;

Permissible yops   *Permissible conds*

AY0	EQ LE	AC
AY1	NE NEG	NOT AC
AF	GT POS	MV
	GE AV	NOT MV
	LT NOT AV	NOT CE

Example: IF GT AF = AF + 1;

Description: Test the optional condition and if true, then increment the source operand by H#0001 and store in the destination location. If the condition is not true then perform a no-operation. Omitting the condition performs the increment operation unconditionally. The source operand is contained in the data register specified in the instruction.

Status Generated:

ASTAT:	7	6	5	4	3	2	1	0
	SS	MV	AQ	AS	AC	AV	AN	AZ
	–	–	–	–	*	*	*	*

AZ   Set if the result equals zero. Cleared otherwise.
AN   Set if the result is negative. Cleared otherwise.
AV   Set if an overflow is generated. Cleared otherwise.
AC   Set if a carry is generated. Cleared otherwise.

Instruction Format:

Conditional ALU/MAC Operation, Instruction Type 9:

23 22 21 20 19	18	17 16 15 14 13 12 11	10 9 8	7 6 5 4	3 2 1 0
0 0 1 0 0	Z	AMF	Yop	Xop	0 0 0 0	COND

AMF specifies the ALU or MAC operation. In this case,
    AMF = 10001 for yop + 1 operation.
Note that the xop field is ignored for the increment operation.

Z:	Destination register	Yop:	Y operand
Xop:	X operand	COND:	condition

12 ALU DECREMENT

Syntax: [IF cond] $\left| \begin{array}{l} \text{AR} \\ \text{AF} \end{array} \right|$ = yop – 1 ;

Permissible yops   *Permissible conds*
AY0   EQ LE   AC
AY1   NE NEG   NOT AC
AF   GT POS   MV
             GE AV   NOT MV
             LT NOT AV   NOT CE

Example: IF EQ AR = AY1 – 1 ;

Description: Test the optional condition and if true, then decrement the source operand by H#0001 and store in the destination location. If the condition is not true then perform a no-operation. Omitting the condition performs the decrement operation unconditionally. The source operand is contained in the data register specified in the instruction.

Status Generated:

ASTAT: 7   6   5   4   3   2   1   0
             SS   MV   AQ   AS   AC   AV   AN   AZ
             –   –   –   –   *   *   *   *

AZ   Set if the result equals zero. Cleared otherwise.
AN   Set if the result is negative. Cleared otherwise.
AV   Set if an overflow is generated. Cleared otherwise.
AC   Set if a carry is generated. Cleared otherwise.

Instruction Format:

Conditional ALU/MAC Operation, Instruction Type 9:

23 22 21 20 19 18	17 16 15 14 13	12 11	10 9 8	7 6 5 4	3 2 1 0
0 0 1 0 0 Z	AMF	Yop	Xop	0 0 0 0	COND

AMF specifies the ALU or MAC operation. In this case,
     AMF = 11000 for yop – 1 operation.
Note that the xop field is ignored for the decrement operation.

Z: Destination register   Yop: Y operand
Xop: X operand   COND: condition

Syntax: DIVS yop , xop ;
              DIVQ xop ;

Permissible xops   *Permissible yops*
AX0 MR2   AY1
AX1 MR1   AF
AR   MR0
     SR1
     SR0

Description: These instructions implement *yop* / *xop*. There are two divide primitives, DIVS and DIVQ. A single precision divide, with a 32-bit numerator and a 16-bit denominator, yielding a 16-bit quotient, executes in 16 cycles. Higher precision divides are also possible.

The division can be either signed or unsigned, but both the numerator and denominator must be the same; both signed or unsigned. The programmer sets up the divide by sorting the upper half of the numerator in any permissible *yop* (AY1 or AF), the lower half of the numerator in AY0, and the denominator in any permissible *xop*. The divide operation is then executed with the divide primitives, DIVS and DIVQ. Repeated execution of DIVQ implements a non-restoring conditional add-subtract division algorithm. At the conclusion of the divide operation the quotient will be in AY0.

To implement a signed divide, first execute the DIVS instruction once, which computes the sign of the quotient. Then execute the DIVQ instruction for as many times as there are bits remaining in the quotient (e.g., for a signed, single-precision divide, execute DIVS once and DIVQ 15 times).

To implement an unsigned divide, first place the upper half of the numerator in AF, then set the AQ bit to zero by manually clearing it in the Arithmetic Status Register, ASTAT. This indicates that the sign of the quotient is positive. Then execute the DIVQ instruction for as many times as there are bits in the quotient (e.g., for an unsigned single-precision divide, execute DIVQ 16 times).

The quotient bit generated on each execution of DIVS and DIVQ is the AQ bit which is written to the ASTAT register at the end of each cycle. The final remainder produced by this algorithm (and left over in the AF register) is not valid and must be corrected if it is needed. For more information, consult the Division Exceptions appendix in this manual.

(instruction continues on next page)

12 ALU
DIVIDE

Status Generated:

ASTAT:	7	6	5	4	3	2	1	0
	SS	MV	AQ	AS	AC	AV	AN	AZ
	−	−	*	−	−	−	−	−

AQ   Loaded with the bit value equal to the AQ bit computed on each cycle from execution of the DIVS or DIVQ instruction.

Instruction Format:

DIVQ, Instruction Type 23:

23	22	21	20	19	18	17	16	15	14	13	12	11	10 9 8	7 6 5 4 3 2 1 0
0	0	0	0	0	0	1	1	1	0	0	0	1	0	Xop	0 0 0 0 0 0 0 0

DIVS, Instruction Type 24:

23	22	21	20	19	18	17	16	15	14	13	12 11	10 9 8	7 6 5 4 3 2 1 0
0	0	0	0	0	1	1	0	0	0	0	Yop	Xop	0 0 0 0 0 0 0 0

Xop: X operand   Yop: Y operand

Syntax:

$$[\ \text{IF cond}] \ \left|\ \begin{array}{c} \text{MR} \\ \text{MF} \end{array}\ \right| \ = \text{xop} * \text{yop} \ \left|\ \begin{array}{c} \text{(SS)} \\ \text{(SU)} \\ \text{(US)} \\ \text{(UU)} \\ \text{(RND)} \end{array}\ \right| \ ;$$

Permissible xops		Permissible yops	Permissible conds		
MX0	AR	MY0	EQ	LE	AC
MX1	SR1	MY1	NE	NEG	NOT AC
MR2	SR0	MF	GT	POS	MV
MR1			GE	AV	NOT MV
MR0			LT	NOT AV	NOT CE

Example: IF EQ MR = MX0 * MF (UU);

Description: Test the optional condition and, if true, then multiply the two source operands and store in the destination location. If the condition is not true perform a no-operation. Omitting the condition performs the multiplication unconditionally. The operands are contained in the data registers specified in the instruction. When MF is the destination operand, only bits 31-16 of the product are stored in MF.

The data format selection field following the two operands specifies whether each respective operand is in Signed (S) or Unsigned (U) format. The *xop* is specified first and *yop* is second. There is no default; one of the data formats must be specified.

If RND (Round) is specified, the MAC multiplies the two source operands, rounds the result to the most significant 24 bits (or rounds bits 31-16 to 16 bits if there is no overflow from the multiply), and stores the result in the destination location. The two multiplication operands *xop* and *yop* are considered to be in twos complement format. All rounding is unbiased. For a discussion of unbiased rounding, see "Rounding Mode" in the "Multiplier/Accumulator" section of this manual.

Status Generated:

ASTAT:	7	6	5	4	3	2	1	0
	SS	MV	AQ	AS	AC	AV	AN	AZ
	–	*	–	–	–	–	–	–

MV   Set on MAC overflow (if any of upper 9 bits of MR are not all one or zero). Cleared otherwise.

(instruction continues on next page)

12 MAC MULTIPLY

Instruction Format:

Conditional ALU/MAC Operation, Instruction Type 9:

23 22 21 20 19	18	17 16 15 14 13 12 11	10 9	8 7	6 5 4 3	2 1 0
0 0 1 0 0	Z	AMF	Yop	Xop	0 0 0 0	COND

AMF: Specifies the ALU or MAC Operation. In this case,

AMF	FUNCTION	Data Format	X-Operand	Y-Operand
0 0 1 0 0	xop * yop	(SS)	Signed	Signed
0 0 1 0 1	xop * yop	(SU)	Signed	Unsigned
0 0 1 1 0	xop * yop	(US)	Unsigned	Signed
0 0 1 1 1	xop * yop	(UU)	Unsigned	Unsigned
0 0 0 0 1	xop * yop	(RND)	Signed	Signed

Z:	Destination register	Yop:	Y operand
Xop:	X operand	COND:	condition

Syntax: [IF cond] | MR | = MR + xop * yop | (SS) | ;
                               | MF | | (SU) |
                                                            | (US) |
                                                            | (UU) |
                                                            | (RND) |

Permissible xops		*Permissible yops*	*Permissible conds*		
MX0	AR	MY0	EQ	LE	AC
MX1	SR1	MY1	NE	NEG	NOT AC
MR2	SR0	MF	GT	POS	MV
MR1			GE	AV	NOT MV
MR0			LT	NOT AV	NOT CE

Example: IF GE MR = MR + MX0 * MY1 (SS) ;

Description: Test the optional condition and, if true, then multiply the two source operands, add the product to the present contents of the MR register, and store the result in the destination location. If the condition is not true then perform a no-operation. Omitting the condition performs the multiply / accumulate unconditionally. The operands are contained in the data registers specified in the instruction. When MF is the destination operand, only bits 31-16 of the 40-bit result are stored in MF.

The data format selection field to the right of the two operands specifies whether each respective operand is in signed (S) or unsigned (U) format. The X operand is specified first and Y operand is second. There is no default. A data format must be specified.

If RND (Round) is specified, the MAC multiplies the two source operands, adds the product to the current contents of the MR register, rounds the result to the most significant 24 bits (or rounds bits 31-16 to the nearest 16 bits if there is no overflow from the multiply/accumulate), and stores the result in the destination location. The two multiplication operands *xop* and *yop* are considered to be in signed twos complement format. All rounding is unbiased. For a discussion of unbiased rounding, see "Rounding Mode" in the "Multiplier/Accumulator" section of this manual.

(instruction continues on next page)

12 MAC
MULTIPLY / ACCUMULATE

Status Generated:

ASTAT:

7	6	5	4	3	2	1	0
SS	MV	AQ	AS	AC	AV	AN	AZ
–	*	–	–	–	–	–	–

MV   Set on MAC overflow (if any of upper 9 bits of MR are not all one or zero). Cleared otherwise.

Instruction Format:

Conditional ALU/MAC Operation, Instruction Type 9:

23 22 21 20 19	18	17 16 15 14 13 12 11	10 9	8 7 6 5	4 3 2 1 0
0 0 1 0 0	Z	AMF	Yop	Xop 0 0 0 0	COND

AMF: Specifies the ALU or MAC Operation. In this case,

AMF	FUNCTION	Data Format	X-Operand	Y-Operand
0 1 0 0 0	MR+xop * yop	(SS)	Signed	Signed
0 1 0 0 1	MR+xop * yop	(SU)	Signed	Unsigned
0 1 0 1 0	MR+xop * yop	(US)	Unsigned	Signed
0 1 0 1 1	MR+xop * yop	(UU)	Unsigned	Unsigned
0 0 0 1 0	MR+xop * yop	(RND)	Signed	Signed

Z:	Destination register	Yop:	Y operand
Xop:	X operand	COND:	condition

278

Syntax:

| [IF cond] | MR
MF | = MR − xop * yop | (SS)
(SU)
(US)
(UU)
(RND) | ; |

Permissible xops		*Permissible yops*	*Permissible conds*		
MX0	AR	MY0	EQ	LE	AC
MX1	SR1	MY1	NE	NEG	NOT AC
MR2	SR0	MF	GT	POS	MV
MR1			GE	AV	NOT MV
MR0			LT	NOT AV	NOT CE

Example: IF LT MR = MR − MX1 * MY0 (SU) ;

Description: Test the optional condition and, if true, then multiply the two source operands, subtract the product from the present contents of the MR register, and store the result in the destination location. If the condition is not true perform a no-operation. Omitting the condition performs the multiply/subtract unconditionally. The operands are contained in the data registers specified in the instruction. When MF is the destination operand, only bits 16-31 of the 40-bit result are stored in MF.

The data format selection field to the right of the two operands specifies whether each respective operand is in signed (S) or unsigned (U) format. The X operand is specified first and Y operand is second. There is no default; a data format must be specified.

If RND (Round) is specified, the MAC multiplies the two source operands, subtracts the product from the current contents of the MR register, rounds the result to the most significant 24 bits (or rounds bits 31-16 to 16 bits if there is no overflow from the multiply/accumulate), and stores the result in the destination location. The two multiplication operands *xop* and *yop* are considered to be in signed twos complement format. All rounding is unbiased. For a discussion of unbiased rounding, see "Rounding Mode" in the "Multiplier/Accumulator" section of this manual.

(instruction continues on next page)

12 MAC
MULTIPLY / SUBTRACT

Status Generated:

ASTAT:

7	6	5	4	3	2	1	0
SS	MV	AQ	AS	AC	AV	AN	AZ
–	*	–	–	–	–	–	–

MV   Set on MAC overflow (if any of the upper 9 bits of MR are not all one or zero). Cleared otherwise.

Instruction Format:

Conditional ALU/MAC Operation, Instruction Type 9:

23 22 21 20 19	18	17 16 15 14 13	12 11	10 9 8	7 6 5 4	3 2 1 0
0 0 1 0 0	Z	AMF	Yop	Xop	0 0 0 0	COND

AMF: Specifies the ALU or MAC Operation. In this case,

AMF	FUNCTION	Data Format	X-Operand	Y-Operand
0 1 1 0 0	MR–xop * yop	(SS)	Signed	Signed
0 1 1 0 1	MR–xop * yop	(SU)	Signed	Unsigned
0 1 1 1 0	MR–xop * yop	(US)	Unsigned	Signed
0 1 1 1 1	MR–xop * yop	(UU)	Unsigned	Unsigned
0 0 0 1 1	MR–xop * yop	(RND)	Signed	Signed

Z:	Destination register	Yop:	Y operand
Xop:	X operand	COND:	condition

Syntax: [IF cond] | MR | = 0 ;
                               | MF |

Permissible conds

EQ	NE	GT	GE	LT
LE	NEG	POS	AV	NOT AV
AC	NOT AC	MV	NOT MV	NOT CE

Example: IF GT MR = 0;

Description: Test the optional condition and, if true, then set the specified register to zero. If the condition is not true perform a no-operation. Omitting the condition performs the clear unconditionally. The entire 40-bit MR or 16-bit MF register is cleared to zero.

Status Generated:

ASTAT:	7	6	5	4	3	2	1	0
	SS	MV	AQ	AS	AC	AV	AN	AZ
	–	0	–	–	–	–	–	–

MV   Always cleared.

Instruction Format:

Conditional ALU/MAC Operation, Instruction Type 9:

23	22	21	20	19	18	17	16	15	14	13	12	11	10	9	8	7	6	5	4	3	2	1	0
0	0	1	0	0	Z		AMF					1	1		0	0	0	0	0	0	0	COND	

AMF: Specifies the ALU or MAC Operation. In this case,
      AMF = 00100 for clear operation.

Note that this instruction is a special case of xop * yop, with yop set to zero.

Z: Destination register   COND: condition

12

MAC
TRANSFER MR

Syntax: [IF cond] $\begin{vmatrix} MR \\ MF \end{vmatrix}$ = MR [(RND)] ;

Permissible conds

EQ	NE	GT	GE	LT
LE	NEG	POS	AV	NOT AV
AC	NOT AC	MV	NOT MV	NOT CE

Example: IF EQ MF = MR (RND);

Description: Test the optional condition and, if true, then perform the MR transfer according to the description below. If the condition is not true then perform a no-operation. Omitting the condition performs the transfer unconditionally.

This instruction actually performs a multiply/accumulate, specifying *yop* = 0 as a multiplicand and adding the zero product to the contents of MR. The MR register may be optionally rounded at the boundary between bits 15 and 16 of the result by specifying the RND option. If MF is specified as the destination, bits 31-16 of the result are stored in MF. If MR is the destination, the entire 40-bit result is stored in MR.

Status Generated:

ASTAT:

7	6	5	4	3	2	1	0
SS	MV	AQ	AS	AC	AV	AN	AZ
–	*	–	–	–	–	–	–

MV   Set on MAC overflow (if any of upper 9 bits of MR are not all one or zero). Cleared otherwise.

Instruction Format:

Conditional ALU/MAC Operation, Instruction Type 9:

23 22 21 20 19 18	17 16 15 14 13 12	11 10 9 8	7 6 5 4	3 2 1 0
0 0 1 0 0 Z	AMF	1 1	0 0 0 0	0 0 0 COND

AMF: Specifies the ALU or MAC Operation. In this case,

AMF = 01000 for Transfer MR operation

Note that this instruction is a special case of MR + xop * yop, with yop set to zero.

Z: Destination register   COND: condition

Syntax: IF MV SAT MR ;

Description: Test the MV (MAC Overflow) bit in the Arithmetic Status
Register (ASTAT), and if set, then saturate the lower-order 32 bits of the
40-bit MR register; if the MV is not set then perform a no-operation.

Saturation of MR is executed with this instruction for one cycle only; MAC
saturation is not a continuous mode that is enabled or disabled. The
saturation instruction is intended to be used at the completion of a series
of multiply/accumulate operations so that temporary overflows do not
cause the accumulator to saturate.

The saturation result depends on the state of MV and on the sign of MR
(the MSB of MR2). The possible results after execution of the saturation
instruction are shown in the table below.

MV	MSB of MR2	MR contents after saturation
0	0	No change
0	1	No change
1	0	00000000 0111111111111111 1111111111111111
1	1	11111111 1000000000000000 0000000000000000

Status Generated: No status bits affected.

Instruction Format:
Saturate MR operation, Instruction Type 25:

23	22	21	20	19	18	17	16	15	14	13	12	11	10	9	8	7	6	5	4	3	2	1	0
0	0	0	0	0	1	0	1	0	0	0	0	0	0	0	0	0	0	0	0	0	0	0	0

12 SHIFTER
ARITHMETIC SHIFT

Syntax: [IF cond] SR = [SR OR] ASHIFT xop $\left|\begin{matrix}(HI)\\(LO)\end{matrix}\right|$;

Permissible xops

SI	AR
SR1	MR2
SR0	MR1
	MR0

Permissible conds

EQ	LE	AC
NE	NEG	NOT AC
GT	POS	MV
GE	AV	NOT MV
LT	NOT AV	NOT CE

Example: IF LT SR = SR OR ASHIFT SI (LO);

Description: Test the optional condition and, if true, then perform the designated arithmetic shift. If the condition is not true then perform a no-operation. Omitting the condition performs the shift unconditionally. The operation arithmetically shifts the bits of the operand by the amount and direction specified in the Shift Code from the SE register. Positive Shift Codes cause a left shift (upshift) and negative codes cause a right shift (downshift).

The shift may be referenced to the upper half of the output field (HI option) or to the lower half (LO option). The shift output may be logically ORed with the present contents of the SR register by selecting the SR OR option.

For ASHIFT with a positive Shift Code (i.e. positive value in SE), the operand is shifted left; with a negative Shift Code (i.e. negative value in SE), the operand is shifted right. The number of positions shifted is the count in the Shift Code. The 32-bit output field is sign-extended to the left (the MSB of the input is replicated to the left), and the output is zero-filled from the right. Bits shifted out of the high order bit in the 32-bit destination field (SR_{31}) are dropped. Bits shifted out of the low order bit in the destination field (SR_0) are dropped.

To shift a double precision number, the same Shift Code is used for both halves of the number. On the first cycle, the upper half of the number is shifted using an ASHIFT with the HI option; on the following cycle, the lower half of the number is shifted using an LSHIFT with the LO and OR options. This prevents sign bit extension of the lower word's MSB.

Status Generated: None affected.

Instruction Format:

Conditional Shift Operation, Instruction Type 16:

23 22 21 20 19 18 17 16 15	14 13 12 11	10 9 8	7 6 5 4	3 2 1 0
0 0 0 0 1 1 1 0 0	SF	Xop	0 0 0 0	COND

SF	*Shifter Function*
0 1 0 0	ASHIFT (HI)
0 1 0 1	ASHIFT (HI, OR)
0 1 1 0	ASHIFT (LO)
0 1 1 1	ASHIFT (LO, OR)

Xop: shifter operand   COND: condition

12 SHIFTER
LOGICAL SHIFT

Syntax: [IF cond] SR = [SR OR] LSHIFT xop $\left|\begin{matrix} \text{(HI)} \\ \text{(LO)} \end{matrix}\right|$;

Permissible xops		*Permissible conds*		
SI	AR	EQ	LE	AC
SR1	MR2	NE	NEG	NOT AC
SR0	MR1	GT	POS	MV
	MR0	GE	AV	NOT MV
		LT	NOT AV	NOT CE

Example: IF GE SR = LSHIFT SI (HI) ;

Description: Test the optional condition and, if true, then perform the designated logical shift. If the condition is not true then perform a no-operation. Omitting the condition performs the shift unconditionally. The operation logically shifts the bits of the operand by the amount and direction specified in the Shift Code from the SE register. Positive Shift Codes cause a left shift (upshift) and negative Codes cause a right shift (downshift).

The shift may be referenced to the upper half of the output field (HI option) or to the lower half (LO option). The shift output may be logically ORed with the present contents of the SR register by selecting the SR OR option.

For LSHIFT with a positive Shift Code, the operand is shifted left; the numbers of positions shifted is the count in the Shift Code. The 32-bit output field is zero-filled from the right. Bits shifted out of the high order bit in the 32-bit destination field (SR_{31}) are dropped.

For LSHIFT with a negative Shift Code, the operand is shifted right; the number of positions shifted is the count in the Shift Code. The 32-bit output field is zero-filled from the left. Bits shifted out of the low order bit in the destination field (SR_0) are dropped.

To shift a double precision number, the same Shift Code is used for both halves of the number. On the first cycle, the upper half of the number is shifted using the HI option; on the following cycle, the lower half of the number is shifted using the LO and OR options.

Status Generated: None affected.

Instruction Format:

Conditional Shift Operation, Instruction Type 16:

23	22	21	20	19	18	17	16	15	14 13 12 11	10 9 8	7 6 5 4	3 2 1 0
0	0	0	0	1	1	1	0	0	SF	Xop	0 0 0 0	COND

SF	*Shifter Function*
0 0 0 0	LSHIFT (HI)
0 0 0 1	LSHIFT (HI, OR)
0 0 1 0	LSHIFT (LO)
0 0 1 1	LSHIFT (LO, OR)

Xop: shifter operand   COND: condition

12 SHIFTER NORMALIZE

Syntax: [IF cond] SR = [SR OR] NORM   xop | (HI) | ;
                                                     | (LO) |

Permissible xops   *Permissible conds*
SI   AR   EQ   LE   AC
SR1   MR2   NE   NEG   NOT AC
SR0   MR1   GT   POS   MV
        MR0   GE   AV   NOT MV
                    LT   NOT AV   NOT CE

Example: SR = NORM SI (HI) ;

Description: Test the optional condition and, if true, then perform the designated normalization. If the condition is not true then perform a no-operation. Omitting the condition performs the normalize unconditionally. The operation arithmetically shifts the input operand to eliminate all but one of the sign bits. The amount of the shift comes from the SE register. The SE register may be loaded with the proper Shift Code to eliminate the redundant sign bits by using the Derive Exponent instruction; the Shift Code loaded will be the negative of the quantity: (the number of sign bits minus one).

The shift may be referenced to the upper half of the output field (HI option) or to the lower half (LO option). The shift output may be logically ORed with the present contents of the SR register by selecting the SR OR option. When the LO reference is selected, the 32-bit output field is zero-filled to the left. Bits shifted out of the high order bit in the 32-bit destination field (SR_{31}) are dropped.

The 32-bit output field is zero-filled from the right. If the exponent of an overflowed ALU result was derived with the HIX modifier, the 32-bit output field is filled from left with the ALU Carry (AC) bit in the Arithmetic Status Register (ASTAT) during a NORM (HI) operation. In this case (SE=1 from the exponent detection on the overflowed ALU value) a downshift occurs.

To normalize a double precision number, the same Shift Code is used for both halves of the number. On the first cycle, the upper half of the number is shifted using the HI option; on the following cycle, the lower half of the number is shifted using the LO and OR options.

Status Generated: None affected.

Instruction Format:
Conditional Shift Operation, Instruction Type 16:

23 22 21 20 19 18 17 16 15	14 13 12 11	10 9 8	7 6 5 4	3 2 1 0
0 0 0 0 1 1 1 0 0	SF	Xop	0 0 0 0	COND

SF	Shifter Function
1 0 0 0	NORM (HI)
1 0 0 1	NORM (HI, OR)
1 0 1 0	NORM (LO)
1 0 1 1	NORM (LO, OR)

Xop: shifter operand   COND: condition

12 SHIFTER
DERIVE EXPONENT

Syntax: [IF cond] SE = EXP   xop $\begin{vmatrix} \text{(HI)} \\ \text{(LO)} \\ \text{(HIX)} \end{vmatrix}$;

Permissible xops		*Permissible conds*		
SI	AR	EQ	LE	AC
SR1	MR2	NE	NEG	NOT AC
SR0	MR1	GT	POS	MV
	MR0	GE	AV	NOT MV
		LT	NOT AV	NOT CE

Example: IF GT SE = EXP MR1 (HI) ;

Description: Test the optional condition and, if true, perform the designated exponent operation. If the condition is not true then perform a no-operation. Omitting the condition performs the exponent operation unconditionally.

The EXP operation derives the effective exponent of the input operand to prepare for the normalization operation (NORM). EXP supplies the source operand to the exponent detector, which generates a Shift Code from the number of leading sign bits in the input operand. The Shift Code, stored in SE at the completion of the EXP instruction, is the effective exponent of the input value. The Shift Code depends on which exponent detector mode is used (HI, HIX, LO).

In the HI mode, the input is interpreted as a single precision signed number, or as the upper half of a double precision signed number. The exponent detector counts the number of leading sign bits in the source operand and stores the resulting Shift Code in SE. The Shift Code will equal the negative of the number of redundant sign bits in the input.

In the HIX mode, the input is interpreted as the result of an add or subtract which may have overflowed. HIX is intended to handle shifting and normalization of results from ALU operations. The HIX mode examines the ALU Overflow bit (AV) in the Arithmetic Status Register: if AV is set, then the effective exponent of the input is +1 (indicating that an ALU overflow occurred before the EXP operation), and +1 is stored in SE. If AV is not set, then HIX performs exactly the same operations as the HI mode.

In the LO mode, the input is interpreted as the lower half of a double precision number. In performing the EXP operation on a double precision number, the higher half of the number must first be processed with EXP in the HI or HIX mode, and then the lower half can be processed with EXP in the LO mode. If the upper half contained a non-sign bit, then the correct Shift Code was generated in the HI or HIX operation and that is the code that is stored in SE. If, however, the upper half was all sign bits, then EXP in the LO mode totals the number of leading sign bits in the double precision word and stores the resulting Shift Code in SE.

Status Generated:

ASTAT:	7	6	5	4	3	2	1	0
	SS	MV	AQ	AS	AC	AV	AN	AZ
	*	–	–	–	–	–	–	–

SS   Set by the MSB of the input for an EXP operation in the HI or HIX mode with AV = 0. Set by the MSB inverted in the HIX mode with AV = 1. Not affected by operations in the LO mode.

Instruction Format:

Conditional Shift Operation, Instruction Type 16:

23	22	21	20	19	18	17	16	15	14 13 12 11	10 9 8	7 6 5 4	3 2 1 0
0	0	0	0	1	1	1	0	0	SF	Xop	0 0 0 0	COND

SF   *Shifter Function*
1 1 0 0   EXP (HI)
1 1 0 1   EXP (HIX)
1 1 1 0   EXP (LO)

Xop: shifter operand   COND: condition

12 SHIFTER
BLOCK EXPONENT ADJUST

Syntax: [IF cond] SB = EXPADJ   xop ;

Permissible xops		*Permissible conds*		
SI	AR	EQ	LE	AC
SR1	MR2	NE	NEG	NOT AC
SR0	MR1	GT	POS	MV
	MR0	GE	AV	NOT MV
		LT	NOT AV	NOT CE

Example: IF GT SB = EXPADJ SI ;

Description: Test the optional condition and, if true, perform the designated exponent operation. If the condition is not true then perform a no-operation. Omitting the condition performs the exponent operation unconditionally. The Block Exponent Adjust operation, when performed on a series of numbers, derives the effective exponent of the number largest in magnitude. This exponent can then be associated with all of the numbers in a block floating point representation.

The Block Exponent Adjust circuitry applies the input operand to the exponent detector to derive its effective exponent. The input must be a signed twos complement number. The exponent detector operates in HI mode (see the EXP instruction, above).

At the start of a block, the SB register should be initialized to –16 to set SB to its minimum value. On each execution of the EXPADJ instruction, the effective exponent of each operand is compared to the current contents of the SB register. If the new exponent is greater than the current SB value, it is written to the SB register, updating it. Therefore, at the end of the block, the SB register will contain the largest exponent found. EXPADJ is only an inspection operation; no actual shifting takes place since the true exponent is not known until all the numbers in the block have been checked. However, the numbers can be shifted at a later time after the true exponent has been derived.

Extended (overflowed) numbers and the lower halves of double precision numbers can not be processed with the Block Exponent Adjust instruction.

Status Generated: Not affected.

Instruction Format:

Conditional Shift Operation, Instruction Type 16:

23	22	21	20	19	18	17	16	15	14 13 12 11	10 9 8	7 6 5 4	3 2 1 0
0	0	0	0	1	1	1	0	0	SF	Xop	0 0 0 0	COND

SF = 1111.

Xop: shifter operand   COND: condition

12 SHIFTER
ARITHMETIC SHIFT IMMEDIATE

Syntax: SR = [SR OR] ASHIFT xop BY <exp> | (HI) | ;
                                                      | (LO) |

Permissible xops   *<exp>*
SI   MR0   Any constant between –128 and 127*
SR1   MR1
SR0   MR2
AR

Example: SR = SR OR ASHIFT SR0 BY 3 (LO);   {do not use "+3"}

Description: Arithmetically shift the bits of the operand by the amount and direction specified by the constant in the exponent field. Positive constants cause a left shift (upshift) and negative constants cause a right shift (downshift). A positive constant must be entered **without** a "+" sign.

The shift may be referenced to the upper half of the output field (HI option) or to the lower half (LO option). The shift output may be logically ORed with the present contents of the SR register by selecting the SR OR option.

For ASHIFT with a positive shift constant the operand is shifted left; with a negative shift constant the operand is shifted right. The 32-bit output field is sign-extended to the left (the MSB of the input is replicated to the left), and the output is zero-filled from the right. Bits shifted out of the high order bit in the 32-bit destination field (SR_{31}) are dropped. Bits shifted out of the low order bit in the destination field (SR_0) are dropped.

To shift a double precision number, the same shift constant is used for both halves of the number. On the first cycle, the upper half of the number is shifted using an ASHIFT with the HI option; on the following cycle, the lower half is shifted using an LSHIFT with the LO and OR options. This prevents sign bit extension of the lower word's MSB.

Status Generated: None affected.

Instruction Format:

Shift Immediate Operation, Instruction Type 15:

23	22	21	20	19	18	17	16	15	14 13 12 11	10 9 8	7 6 5 4 3 2 1 0
0	0	0	0	1	1	1	1	0	SF	Xop	<exp>

SF	Shifter Function
0 1 0 0	ASHIFT (HI)
0 1 0 1	ASHIFT (HI, OR)
0 1 1 0	ASHIFT (LO)
0 1 1 1	ASHIFT (LO, OR)

Xop: Shifter Operand   <exp>:   8-bit signed shift value

* See Table 2.4 in Chapter 2.

12 SHIFTER
LOGICAL SHIFT IMMEDIATE

Syntax: SR = [SR OR] LSHIFT xop BY <exp> | (HI) | ;
                                                      | (LO) |

Permissible xops   *<exp>*
SI   MR0   Any constant between −128 and 127*
SR1   MR1
SR0   MR2
AR

Example: SR = LSHIFT SR1 BY −6 (HI) ;

Description: Logically shifts the bits of the operand by the amount and direction specified by the constant in the exponent field. Positive constants cause a left shift (upshift); negative constants cause a right shift (downshift). A positive constant must be entered **without** a "+" sign.

The shift may be referenced to the upper half of the output field (HI option) or to the lower half (LO option). The shift output may be logically ORed with the contents of the SR register by selecting the SR OR option.

For LSHIFT with a positive shift constant, the operand is shifted left. The 32-bit output field is zero-filled to the left and from the right. Bits shifted out of the high order bit in the 32-bit destination field (SR_{31}) are dropped. For LSHIFT with a negative shift constant, the operand is shifted right. The 32-bit output field is zero-filled from the left and to the right. Bits shifted out of the low order bit are dropped.

To shift a double precision number, the same shift constant is used for both parts of the number. On the first cycle, the upper half of the number is shifted using the HI option; on the following cycle, the lower half is shifted using the LO and OR options.

Status Generated: None affected.

Instruction Format:
Shift Immediate Operation, Instruction Type 15:

23 22 21 20 19 18 17 16 15	14 13 12 11	10 9 8	7 6 5 4 3 2 1 0
0 0 0 0 1 1 1 1 0	SF	Xop	<exp>

SF   *Shifter Function*
0 0 0 0   LSHIFT (HI)   Xop: Shifter Operand
0 0 0 1   LSHIFT (HI, OR)
0 0 1 0   LSHIFT (LO)   <exp>: 8-bit signed shift value
0 0 1 1   LSHIFT (LO, OR)

* See Table 2.4 in Chapter 2.

Syntax: reg = reg ;

Permissible registers

AX0	MX0	SI	SB	CNTR	
AX1	MX1	SE	PX	OWRCNTR(*write only*)	
AY0	MY0	SR1	ASTAT	RX0	
AY1	MY1	SR0	MSTAT	RX1	(*not ADSP-2100 registers*)
AR	MR2	I0-I7	SSTAT(*read only*)	TX0	
	MR1	M0-M7	IMASK	TX1	
	MR0	L0-L7	ICNTL	IFC(*write only*)	

Example: I7 = AR;

Description: Move the contents of the source to the destination location. The contents of the source are always right-justified in the destination location after the move.

When transferring a smaller register to a larger register (e.g., an 8-bit register to a 16-bit register), the value stored in the destination is either sign-extended to the left if the source is a signed value, or zero-filled to the left if the source is an unsigned value. The unsigned registers which (when used as the source) cause the value stored in the destination to be zero-filled to the left are: I0 through I7, L0 through L7, CNTR, PX, ASTAT, MSTAT, SSTAT, IMASK, and ICNTL. All other registers cause sign-extension to the left.

When transferring a larger register to a smaller register (e.g., a 16-bit register to a 14-bit register), the value stored in the destination is right-justified (bit 0 maps to bit 0) and the higher-order bits are dropped.

Note that whenever MR1 is loaded with data, it is sign-extended into MR2.

Status Generated: None affected.

(instruction continues on next page)

12 MOVE
REGISTER MOVE

Instruction Format:
Internal Data Move, Instruction Type 17:

23	22	21	20	19	18	17	16	15	14	13	12	11 10	9 8	7 6 5 4	3 2 1 0
0	0	0	0	1	1	0	1	0	0	0	0	DST RGP	SRC RGP	DEST REG	SOURCE REG

SRC RGP (Source Register Group) and SOURCE REG (Source Register) select the source register according to the Register Selection Table (see Appendix A).

DST RGP (Destination Register Group) and DEST REG (Destination Register) select the destination register according to the Register Selection Table (see Appendix A).

Syntax: reg = <data> ;
             dreg = <data> ;

data: <constant>
          '%' <symbol>
          '^' <symbol>

Permissible registers

dregs (*Instruction Type 6*)			regs (*Instruction Type 7*)	
(16-bit load)			(*maximum 14-bit load*)	
AX0	MX0	SI	SB	CNTR
AX1	MX1	SE	PX	OWRCNTR (*write only*)
AY0	MY0	SR1	ASTAT	RX0 (*not ADSP-2100 registers*)
AY1	MY1	SR0	MSTAT	RX1
AR	MR2		IMASK	TX0
	MR1		ICNTL	TX1
	MR0		I0-I7	IFC(*write only*)
			M0-M7	
			L0-L7	

Example: I0 = ^*data_buffer*;
             L0=%*data_buffer*;

Description: Move the data value specified to the destination location.
The data may be a constant, or any symbol referenced with the "length of"
(%) or "pointer to" (^) operators. The data value is contained in the
instruction word, with 16 bits for data register loads and up to 14 bits for
other register loads. The value is always right-justified in the destination
location after the load (bit 0 maps to bit 0). When a value of length less than
the length of the destination is moved, it is sign-extended to the left to fill
the destination width.

Note that whenever MR1 is loaded with data, it is sign-extended into MR2.

For this instruction only, the RX and TX registers may be loaded with a
maximum of 14 bits of data (although the registers themselves are 16 bits
wide). To load these registers with 16-bit data, use the register-to-register
move instruction (page 297) or the data memory-to-register move
instruction with direct addressing (page 301).

Status Generated: None affected.

(instruction continues on next page)

12 MOVE
LOAD REGISTER IMMEDIATE

Instruction Format :
Load Data Register Immediate, Instruction Type 6:

23 22 21 20	19 18 17 16 15 14 13 12 11 10 9 8 7 6 5 4	3 2 1 0
0 1 0 0	DATA	DREG

DATA contains the immediate value to be loaded into the Data Register destination location. The data is right-justified in the field, so the value loaded into an N-bit destination register is contained in the lower-order N bits of the DATA field.

DREG selects the destination Data Register for the immediate data value. One of the 16 Data Registers is selected according to the DREG Selection Table (see Appendix A).

Load Non-Data Register Immediate Instruction Type 7:

23 22 21 20	19 18	17 16 15 14 13 12 11 10 9 8 7 6 5 4	3 2 1 0
0 0 1 1	RGP	DATA	REG

DATA contains the immediate value to be loaded into the Non-Data Register destination location. The data is right-justified in the field, so the value loaded into an N-bit destination register is contained in the lower-order N bits of the DATA field.

RGP (Register Group) and REG (Register) select the destination register according to the Register Selection Table (see Appendix A).

300

Syntax: reg = DM (<addr>) ;

Permissible registers

AX0	MX0	SI	SB	CNTR	
AX1	MX1	SE	PX	OWRCNTR (*write only*)	
AY0	MY0	SR1	ASTAT	RX0	
AY1	MY1	SR0	MSTAT	RX1	(*not ADSP-2100 registers*)
AR	MR2	I0-I7		TX0	
	MR1	M0-M7	IMASK	TX1	
	MR0	L0-L7	ICNTL	IFC(*write only*)	

Example: SI = DM(*ad_port0*);

Description: The Read instruction moves the contents of the data memory location to the destination register. The addressing mode is direct addressing (designated by an immediate address value or by a label). The data memory address is stored directly in the instruction word as a full 14-bit field. The contents of the source are always right-justified in the destination register after the read (bit 0 maps to bit 0).

Note that whenever MR1 is loaded with data, it is sign-extended into MR2.

Status Generated: None affected.

Instruction Format:
Data Memory Read (Direct Address), Instruction Type 3:

23 22 21 20	19 18	17 16 15 14 13 12 11 10 9 8 7 6 5 4 3	2 1 0
1 0 0 0	RGP	ADDR	REG

ADDR contains the direct address to the source location in Data Memory.

RGP (Register Group) and REG (Register) select the destination register according to the Register Selection Table (see Appendix A).

12 MOVE
DATA MEMORY READ (Indirect Address)

Syntax: dreg = DM (

I0	,	M0	) ;
I1		M1	
I2		M2	
I3		M3	
I4		M4	
I5		M5	
I6		M6	
I7		M7	

Permissible dregs

AX0	MX0	SI
AX1	MX1	SE
AY0	MY0	SR1
AY1	MY1	SR0
AR	MR2	
	MR1	
	MR0	

Example: AY0 = DM (I3, M1);

Description: The Data Memory Read Indirect instruction moves the contents of the data memory location to the destination register. The addressing mode is register indirect with post-modify. **For linear (i.e. non-circular) indirect addressing, the L register corresponding to the I register used must be set to zero.** The contents of the source are always right-justified in the destination register after the read (bit 0 maps to bit 0).

Status Generated: None affected.

Instruction Format:
ALU / MAC Operation with Data Memory Read, Instruction Type 4:

23	22	21	20	19	18	17 16 15 14 13	12	11	10	9	8	7 6 5 4	3 2	1 0
0	1	1	G	0	0	AMF	0	0	0	0	0	DREG	I	M

AMF specifies the ALU or MAC operation to be performed in parallel with the Data Memory Read. In this case, AMF = 00000, indicating a no-operation for the ALU/MAC function.

DREG selects the destination Data Register . One of the 16 Data Registers is selected according to the DREG Selection Table (see Appendix A).

G specifies which Data Address Generator the I and M registers are selected from. These registers must be from the same DAG as separated by the gray bar above. I specifies the indirect address pointer (I register). M specifies the modify register (M register).

Syntax: dreg = PM (| I4 | , | M4 |) ;
                           | I5 | | M5 |
                           | I6 | | M6 |
                           | I7 | | M7 |

Permissible dregs

AX0	MX0	SI
AX1	MX1	SE
AY0	MY0	SR1
AY1	MY1	SR0
AR	MR2	
	MR1	
	MR0	

Example: MX1 = PM (I6, M5);

Description: The Program Memory Read Indirect instruction moves the contents of the program memory location to the destination register. The addressing mode is register indirect with post-modify. **For linear (i.e. non-circular) indirect addressing, the L register corresponding to the I register used must be set to zero.** The 16 most significant bits of the Program Memory Data bus (PMD_{23-8}) are loaded into the destination register, with bit PMD_8 lining up with bit 0 of the destination register (right-justification). If the destination register is less than 16 bits wide, the most significant bits are dropped. Bits PMD_{7-0} are always loaded into the PX register. You may ignore these bits or read them out on a subsequent cycle.

Status Generated: None affected

Instruction Format:

ALU / MAC Operation with Program Memory Read, Instruction Type 5:

23	22	21	20	19	18	17 16 15 14 13 12	11	10	9	8	7 6 5 4	3 2	1 0
0	1	0	1	0	0	AMF	0	0	0	0	DREG	I	M

AMF specifies the ALU or MAC operation to be performed in parallel with the Data Memory Read. In this case, AMF = 00000, indicating a no-operation for the ALU/MAC function.

DREG selects the destination Data Register. One of the 16 Data Registers is selected according to the Register Selection Table (see Appendix A).

I specifies the indirect address pointer (I register). M specifies the modify register (M register).

12

MOVE
DATA MEMORY WRITE (Direct Address)

Syntax: DM (<addr>) = reg ;

Permissible registers

AX0	MX0	SI	SB	CNTR
AX1	MX1	SE	PX	RX0
AY0	MY0	SR1	ASTAT	RX1 *(not ADSP-2100 registers)*
AY1	MY1	SR0	MSTAT	TX0
AR	MR2	I0-I7	SSTAT*(read only)*	TX1
	MR1	M0-M7	IMASK	
	MR0	L0-L7	ICNTL	

Example: DM (*cntl_port0*) = AR;

Description: Moves the contents of the source register to the data memory location specified in the instruction word. The addressing mode is direct addressing (designated by an immediate address value or by a label). The data memory address is stored directly in the instruction word as a full 14-bit field. Whenever a register less than 16 bits in length is written to memory, the value written is either sign-extended to the left if the source is a signed value, or zero-filled to the left if the source is an unsigned value. The unsigned registers which are zero-filled to the left are: I0 through I7, L0 through L7, CNTR, PX, ASTAT, MSTAT, SSTAT, IMASK, and ICNTL. All other registers are sign-extended to the left.

The contents of the source are always right-justified in the destination location after the write (bit 0 maps to bit 0).

Note that whenever MR1 is loaded with data, it is sign-extended into MR2.

Status Generated: None affected.

Instruction Format:
Data Memory Read (Direct Address), Instruction Type 3:

23 22 21 20	19 18 17 16 15 14 13 12 11 10 9 8 7 6 5 4 3	3 2 1 0
1 0 0 1	RGP	ADDR	REG

ADDR contains the direct address of the destination location in Data Memory.

RGP (Register Group) and REG (Register) select the source register according to the Register Selection Table (see Appendix A).

304

Syntax:

DM (
I0	,	M0	) =	dreg	;
I1		M1		\<data\>	
I2		M2
I3		M3

I4	M4
I5	M5
I6	M6
I7	M7

data: \<constant\>
'%' \<symbol\>
'^' \<symbol\>

Permissible dregs

AX0	MX0	SI
AX1	MX1	SE
AY0	MY0	SR1
AY1	MY1	SR0
AR	MR2	
	MR1	
	MR0	

Example: DM (I2, M0) = MR1;

Description: The Data Memory Write Indirect instruction moves the contents of the source to the data memory location specified in the instruction word. The immediate data may be a constant or any symbol referenced with the "length of" (%) or "pointer to" (^) operators.

The addressing mode is register indirect with post-modify. **For linear (i.e. non-circular) indirect addressing, the L register corresponding to the I register used must be set to zero.** When a register of less than 16 bits is written to memory, the value written is sign-extended to form a 16-bit value. The contents of the source are always right-justified in the destination location after the write (bit 0 maps to bit 0).

Status Generated: None affected.

(instruction continues on next page)

305

12 MOVE
DATA MEMORY WRITE (Indirect Address)

Instruction Format:
ALU / MAC Operation with Data Memory Write, Instruction Type 4:

23 22 21	20	19 18	17 16 15 14 13	12 11 10 9 8	7 6 5 4	3 2	1	0
0 1 1	G	1 0	AMF	0 0 0 0 0	DREG		I	M

Data Memory Write, Immediate Data, Instruction Type 2:

23 22 21	20	19 18 17 16 15 14 13 12 11 10 9 8 7 6 5 4 3 2	1	0
1 0 1	G	Data	I	M

AMF specifies the ALU or MAC operation to be performed in parallel with the Data Memory Write. In this case, AMF = 00000, indicating a no-operation for the ALU / MAC function.

Data represents the actual 16-bit value.

DREG selects the source Data Register. One of the 16 Data Registers is selected according to the Register Selection Table (see Appendix A).

G specifies which Data Address Generator the I and M registers are selected from. These registers must be from the same DAG as separated by the gray bar in the Syntax description above. I specifies the indirect address pointer (I register). M specifies the modify register (M register).

306

Syntax: PM (| I4 | , | M4 |) = dreg ;
                      | I5 | | M5 |
                      | I6 | | M6 |
                      | I7 | | M7 |

Permissible dregs
AX0	MX0	SI
AX1	MX1	SE
AY0	MY0	SR1
AY1	MY1	SR0
AR	MR2	
	MR1	
	MR0	

Example: PM (I6, M5) = AR;

Description: The Program Memory Write Indirect instruction moves the contents of the source to the program memory location specified in the instruction word. The addressing mode is register indirect with post-modify. **For linear (i.e. non-circular) indirect addressing, the L register corresponding to the I register used must be set to zero.** The 16 most significant bits of the Program Memory Data bus (PMD_{23-8}) are loaded from the source register, with bit PMD_8 aligned with bit 0 of the source register (right justification). The 8 least significant bits of the Program Memory Data bus (PMD_{7-0}) are loaded from the PX register. Whenever a source register of length less than 16 bits is written to memory, the value written is sign-extended to form a 16-bit value.

Status Generated: None affected.

Instruction Format:
ALU / MAC Operation with Program Memory Write, Instruction Type 5 (see Appendix A), as shown below:

23 22 21 20 19 18	17 16 15 14 13	12 11 10 9 8	7 6 5 4	3 2	1 0
0 1 0 1 1 0	AMF	0 0 0 0 0	DREG	I	M

AMF specifies the ALU or MAC operation to be performed in parallel with the Program Memory Write. In this case, AMF = 00000, indicating a no-operation for the ALU / MAC function.

DREG selects the source Data Register. One of the 16 Data Registers is selected according to the Register Selection Table (see Appendix A).

I specifies the indirect address pointer (I register). M specifies the modify register (M register).

12 PROGRAM FLOW
JUMP

Syntax: [IF cond] JUMP   (I4)   ;
                                     (I5)
                                     (I6)
                                     (I7)
                                     <addr>

Permissible conds

EQ	NE	GT	GE	LT
LE	NEG	POS	AV	NOT AV
AC	NOT AC	MV	NOT MV	NOT CE

Example: IF NOT CE JUMP *top_loop*;   {CNTR is decremented}

Description: Test the optional condition and, if true, perform the specified jump. If the condition is not true then perform a no-operation. Omitting the condition performs the jump unconditionally. The JUMP instruction causes program execution to continue at the effective address specified by the instruction. The addressing mode may be direct or register indirect.

For direct addressing (using an immediate address value or a label), the program address is stored directly in the instruction word as a full 14-bit field. For register indirect jumps, the selected I register provides the address; it is not post-modified in this case.

If JUMP is the last instruction inside a DO UNTIL loop, you must ensure that the loop stacks are properly handled. If NOT CE is used as the condition, execution of the JUMP instruction decrements the processor's counter (CNTR register).

Status Generated: None affected.

Instruction Field:
Conditional JUMP Direct Instruction Type 10:

23	22	21	20	19	18	17	16	15	14	13	12	11	10	9	8	7	6	5	4	3	2	1	0
0	0	0	1	1	0						ADDR									COND			

Conditional JUMP Indirect Instruction Type 19:

23	22	21	20	19	18	17	16	15	14	13	12	11	10	9	8	7	6	5	4	3	2	1	0
0	0	0	0	1	0	1	1	0	0	0	0	0	0	0	0	0	I	0	0	COND			

I specifies the I register (Indirect Address Pointer).

ADDR: immediate jump address   COND: condition

Syntax: [IF cond] CALL   (I4)   ;
                                    (I5)
                                    (I6)
                                    (I7)
                                    \<addr\>

Permissible conds

EQ	NE	GT	GE	LT
LE	NEG	POS	AV	NOT AV
AC	NOT AC	MV	NOT MV	NOT CE

Example: IF AV CALL *scale_down*;

Description: Test the optional condition and, if true, then perform the specified call. If the condition is not true then perform a no-operation. Omitting the condition performs the call unconditionally. The CALL instruction is intended for calling subroutines. CALL pushes the PC stack with the return address and causes program execution to continue at the effective address specified by the instruction. The addressing modes available for the CALL instruction are direct or register indirect.

For direct addressing (using an immediate address value or a label), the program address is stored directly in the instruction word as a full 14-bit field. For register indirect jumps, the selected I register provides the address; it is not post-modified in this case.

If CALL is the last instruction inside a DO UNTIL loop, you must ensure that the loop stacks are properly handled.

Status Generated: None affected.

Instruction Field:
Conditional JUMP Direct Instruction Type 10:

23	22	21	20	19	18	17 16 15 14 13 12 11 10 9 8 7 6 5	4 3 2 1 0
0	0	0	1	1	1	ADDR	COND

Conditional JUMP Indirect Instruction Type 19:

23	22	21	20	19	18	17	16	15	14	13	12	11	10	9	8	7 6 5	4	3 2	1 0	COND
0	0	0	0	1	0	1	1	0	0	0	0	0	0	0	0	0	I	0 1	COND	

I specifies the I register (Indirect Address Pointer).

ADDR: immediate jump address   COND: condition

12 PROGRAM FLOW
JUMP or CALL ON FLAG IN PIN
(not ADSP-2100 instruction)

Syntax: IF | FLAG_IN | | JUMP | | <addr> | ;
                       | NOT FLAG_IN | | CALL |

Example: IF FLAG_IN JUMP *service_proc_three*;

Description: Test the condition of the FI pin of the processor and, if set to one, perform the specified jump or call. If FI is zero then perform a no-operation. Omitting the flag in condition reduces the instruction to a standard JUMP or CALL.

The JUMP instruction causes program execution to continue at the address specified by the instruction. The addressing mode for the JUMP on FI must be direct.

The CALL instruction is intended for calling subroutines. CALL pushes the PC stack with the return address and causes program execution to continue at the address specified by the instruction. The addressing mode for the CALL on FI must be direct.

If JUMP or CALL is the last instruction inside a DO UNTIL loop, you must ensure that the loop stacks are properly handled.

For direct addressing (using an immediate address value or a label), the program address is stored directly in the instruction word as a full 14-bit field.

Status Generated: None affected.

Instruction Field:
Conditional JUMP or CALL on Flag In Direct Instruction Type 27:

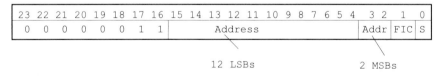

23 22 21 20 19 18 17 16	15 14 13 12 11 10 9 8 7 6 5 4	3 2	1	0
0 0 0 0 0 0 1 1	Address	Addr	FIC	S

                              12 LSBs   2 MSBs

S: specifies JUMP (0) or CALL (1)   FIC: latched state of FI pin

Syntax: [IF cond] | SET | FLAG_OUT | [, ...] ;
	RESET	FL0	*(ADSP-2111 &*
	TOGGLE	FL1	*ADSP21msp50 only)*
		FL2	

Example: IF MV SET FLAG_OUT, RESET FL1;

Description: Evaluate the optional condition and if true, set to one, reset to zero, or toggle the state of the specified flag output pin(s). Otherwise perform a no-operation and continue with the next instruction. Omitting the condition performs the operation unconditionally. Multiple flags may be modified by including multiple clauses, separated by commas, in a single instruction. This instruction does not directly alter the flow of your program—it is provided to signal external devices.

The following table shows which flag outputs are present on each ADSP-21xx processor:

processor	*flag pin(s)*
ADSP-2100	none
ADSP-2101	FO
ADSP-2105	FO
ADSP-2111	FO, FL0, FL1, FL2
ADSP-21msp50	FO, FL0, FL1, FL2

(Note that the FO pin is specified by "FLAG_OUT" in the instruction syntax.)

Status Generated: None affected.

Instruction Field:
Flag Out Mode Control Instruction Type 28:

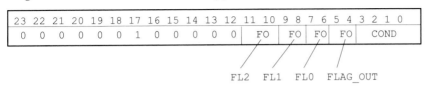

FO: Operation to perform
    on flag output pin

COND: Condition code

12 PROGRAM FLOW
RTS

Syntax: [IF cond]   RTS ;

Permissible conds

EQ	NE	GT	GE	LT
LE	NEG	POS	AV	NOT AV
AC	NOT AC	MV	NOT MV	NOT CE

Example: IF LE RTS ;

Description: Test the optional condition and, if true, then perform the specified return. If the condition is not true then perform a no-operation. Omitting the condition performs the return unconditionally. RTS executes a program return from a subroutine. The address on top of the PC stack is popped and is used as the return address. The PC stack is the only stack popped.

If RTS is the last instruction inside a DO UNTIL loop, you must ensure that the loop stacks are properly handled.

Status Generated: None affected.

Instruction Field:
Conditional Return, Instruction Type 20:

23	22	21	20	19	18	17	16	15	14	13	12	11	10	9	8	7	6	5	4	3	2	1	0
0	0	0	0	1	0	1	0	0	0	0	0	0	0	0	0	0	0	0	0		COND		

COND: condition

Syntax: [IF cond]   RTI ;

Permissible conds

EQ	NE	GT	GE	LT
LE	NEG	POS	AV	NOT AV
AC	NOT AC	MV	NOT MV	NOT CE

Example: IF MV RTI ;

Description: Test the optional condition and, if true, then perform the specified return. If the condition is not true then perform a no-operation. Omitting the condition performs the return unconditionally. RTI executes a program return from an interrupt service routine. The address on top of the PC stack is popped and is used as the return address. The value on top of the status stack is also popped, and is loaded into the arithmetic status (ASTAT), mode status (MSTAT) and the interrupt mask (IMASK) registers.

If RTI is the last instruction inside a DO UNTIL loop, you must ensure that the loop stacks are properly handled.

Status Generated: None affected.

Instruction Field:
Conditional Return, Instruction Type 20:

23	22	21	20	19	18	17	16	15	14	13	12	11	10	9	8	7	6	5	4	3	2	1	0
0	0	0	0	1	0	1	0	0	0	0	0	0	0	0	0	0	0	0	1		COND		

COND: condition

Syntax: DO   <addr>   [UNTIL term] ;

Permissible terms

EQ	NE	GT	GE	LT	FOREVER
LE	NEG	POS	AV	NOT AV	
AC	NOT AC	MV	NOT MV	CE	

Example: DO *loop_label* UNTIL CE ;   {CNTR is decremented
                                                                each pass through loop}

Description: DO UNTIL sets up looping circuitry for zero-overhead looping. The program loop begins at the program instruction immediately following the DO instruction, ends at the address designated in the instruction and repeats execution until the specified termination condition is met (if one is specified) or repeats in an infinite loop (if none is specified). The termination condition is tested during execution of the last instruction in the loop, the status having been generated upon completion of the previous instruction. The address (<addr>) of the last instruction in the loop is stored directly in the instruction word.

If CE is used for the termination condition, the processor's counter (CNTR register) is decremented once for each pass through the loop.

When the DO instruction is executed, the address of the last instruction is pushed onto the loop stack along with the termination condition and the current program counter value plus 1 is pushed onto the PC stack.

Any nesting of DO loops continues the process of pushing the loop and PC stacks, up to the limit of the loop stack size (4 levels of loop nesting) or of the PC stack size (16 levels for subroutines plus interrupts plus loops). With either or both the loop or PC stacks full, a further attempt to perform the DO instruction will set the appropriate stack overflow bit and will perform a no-operation.

Status Generated:
ASTAT: Not affected.

SSTAT:	7	6	5	4	3	2	1	0
	LSO	LSE	SSO	SSE	CSO	CSE	PSO	PSE
	*	0	–	–	–	–	*	0

LSO   Loop Stack Overflow: set if the loop stack overflows; otherwise not affected.

LSE   Loop Stack Empty: always cleared (indicating loop stack not empty)

PSO   PC Stack Overflow: set if the PC stack overflows; otherwise not affected.

PSE   PC Stack Empty: always cleared (indicating PC stack not empty)

Instruction Format:

Do Until, Instruction Type 11:

23 22 21 20 19 18	17 16 15 14 13 12 11 10 9 8 7 6 5 4	3 2 1 0
0 0 0 1 0 1	Addr	TERM

ADDR specifies the address of the last instruction in the loop. In the Instruction Syntax, this field may be a program label or an immediate address value.

TERM specifies the termination condition, as shown below:

TERM	Syntax	Condition Tested
0 0 0 0	NE	Not Equal to Zero
0 0 0 1	EQ	Equal Zero
0 0 1 0	LE	Less Than or Equal to Zero
0 0 1 1	GT	Greater Than Zero
0 1 0 0	GE	Greater Than or Equal to Zero
0 1 0 1	LT	Less Than Zero
0 1 1 0	NOT AV	Not ALU Overflow
0 1 1 1	AV	ALU Overflow
1 0 0 0	NOT AC	Not ALU Carry
1 0 0 1	AC	ALU Carry
1 0 1 0	POS	X Input Sign Positive
1 0 1 1	NEG	X Input Sign Negative
1 1 0 0	NOT MV	Not MAC Overflow
1 1 0 1	MV	MAC Overflow
1 1 1 0	CE	Counter Expired
1 1 1 1	FOREVER	Always

12 PROGRAM FLOW
IDLE
(not ADSP-2100 instruction)

Syntax: IDLE ;

Description: On an ADSP-21xx processor, IDLE waits indefinitely in a
low-power state, waiting for interrupts. When an interrupt occurs it is
serviced and execution continues with the instruction following IDLE.
Typically this next instruction will be a JUMP back to IDLE, implementing
a low-power standby loop. (Note the restrictions on JUMP or IDLE as the
last instruction in a DO UNTIL loop, detailed in Chapter 3.)

Autobuffering continues during IDLE without affecting the idle state.

Status Generated: None affected.

Instruction Field:
Idle Instruction Type 31:

23	22	21	20	19	18	17	16	15	14	13	12	11	10	9	8	7	6	5	4	3	2	1	0
0	0	0	0	0	0	1	0	1	0	0	0	0	0	0	0	0	0	0	0	0	0	0	0

Syntax: [IF cond]   TRAP ;

Permissible conds

EQ	NE	GT	GE	LT
LE	NEG	POS	AV	NOT AV
AC	NOT AC	MV	NOT MV	CE

Example: IF AV TRAP ;

Description: Test the optional condition and, if true, then perform a trap. If the condition is not true then perform a no-operation. Omitting the condition performs the trap unconditionally. TRAP halts the ADSP-2100 and asserts the external TRAP signal. The processor remains halted until HALT is asserted and then released.

The TRAP instruction executes as a NOP on all ADSP-21xx processors other than the ADSP-2100.

The ADSP-2100A Emulator uses the TRAP instruction to implement software breakpoints. When any software breakpoints are set in the emulator, the TRAP instruction does not execute as described above (the TRAP–HALT handshaking is disabled). If your system is designed to utilize this handshaking, you cannot set any software breakpoints in the emulator—hardware event triggering must be used instead. Refer to the section "TRAP Instruction" of Chapter 7, Control and Debug, in the *ADSP-2100A Emulator Manual*.

Status Generated: None affected.

Instruction Format:
Conditional Trap, Instruction Type 22:

23	22	21	20	19	18	17	16	15	14	13	12	11	10	9	8	7	6	5	4	3	2	1	0
0	0	0	0	1	0	0	0	x	x	x	x	x	x	x	x	x	x		COND				

COND: condition

12 MISC
STACK CONTROL

Syntax: $\left[\begin{array}{c}\text{PUSH} \\ \text{POP}\end{array}\Big| \text{STS}\right]$ [, POP CNTR] [, POP PC] [, POP LOOP] ;

Example: POP CNTR, POP PC, POP LOOP;

Description: Stack Control pushes or pops the designated stack(s). The entire instruction executes in one cycle regardless of how many stacks are specified.

The PUSH STS (Push Status Stack) instruction increments the status stack pointer by one to point to the next available status stack location; and pushes the arithmetic status (ASTAT), mode status (MSTAT), and interrupt mask register (IMASK) onto the processor's status stack. Note that the PUSH STS operation is executed automatically whenever an interrupt service routine is entered.

Any POP pops the value on the top of the designated stack and decrements the same stack pointer to point to the next lowest location in the stack. POP STS causes the arithmetic status (ASTAT), mode status (MSTAT), and interrupt mask (IMASK) to be popped into these same registers. This also happens automatically whenever a return from interrupt (RTI) is executed.

POP CNTR causes the counter stack to be popped into the down counter. When the loop stack or PC stack is popped (with POP LOOP or POP PC, respectively), the information is lost. Returning from an interrupt (RTI) or subroutine (RTS) also pops the PC stack automatically.

Syntax: TOPPCSTACK=reg;
reg=TOPPCSTACK;

To retain the value popped from the PC stack, use the instruction "reg = TOPPCSTACK". Any processor register can be used, for example:

```
AX0 = TOPPCSTACK;
```

To push a specific value onto the PC stack, use the instruction "TOPPCSTACK = reg". For example:

```
TOPPCSTACK = AX0;
```

Status Generated:

SSTAT:	7	6	5	4	3	2	1	0
	LSO	LSE	SSO	SSE	CSO	CSE	PSO	PSE
	–	*	*	*	–	*	–	*

PSE   PC Stack Empty: set if a pop results in an empty program counter stack; cleared otherwise.

CSE   Counter Stack Empty: set if a pop results in an empty counter stack; cleared otherwise.

SSE   Status Stack Empty: for PUSH STS, this bit is always cleared (indicating status stack not empty). For POP STS, SSE is set if the pop results in an empty status stack; cleared otherwise.

SSO   Status Stack Overflow: for PUSH STS set if the status stack overflows; otherwise not affected.

LSE   Loop Stack Empty: set if a pop results in an empty loop stack; cleared otherwise.

Note that once any Stack Overflow occurs, the corresponding stack overflow bit is set in SSTAT, and this bit stays set indicating there has been loss of information. Once set, the stack overflow bit can only be cleared by resetting the processor.

Instruction Format:

Stack Control, Instruction Type 26:

23	22	21	20	19	18	17	16	15	14	13	12	11	10	9	8	7	6	5	4	3	2	1	0
0	0	0	0	0	1	0	0	0	0	0	0	0	0	0	0	0	0	0	Pp	Lp	Cp	Spp	

Pp: PC Stack Control   Lp: Loop Stack Control
Cp: Counter Stack Control   Spp: Status Stack Control

TOPPCSTACK=reg
Internal Data Move, Instruction Type 17:

23	22	21	20	19	18	17	16	15	14	13	12	11	10	9	8	7	6	5	4	3	2	1	0
0	0	0	0	1	1	0	1	0	0	0	0	1	1	SRC RGP		1	1	1	1	SOURCE REG			

SRC RGP (Source Register Group) and SOURCE REG (Source Register) select the source register according to the Register Selection Table (see Appendix A).

reg=TOPPCSTACK
Internal Data Move, Instruction Type 17:

23	22	21	20	19	18	17	16	15	14	13	12	11	10	9	8	7	6	5	4	3	2	1	0
0	0	0	0	1	1	0	1	0	0	0	0	DST RGP		1	1	DEST REG		1	1	1	1		

DST RGP (Destination Register Group) and DEST REG (Destination Register) select the destination register according to the Register Selection Table (see Appendix A).

12 MISC
MODE CONTROL

Syntax: | ENA | | BIT_REV | [, ...] ;
| DIS | | AV_LATCH
| | | AR_SAT
| | | SEC_REG
| | | G_MODE
| | | M_MODE   *(not ADSP-2100 modes)*
| | | TIMER

Example: DIS AR_SAT, ENA M_MODE;

Description: Enables (ENA) or disables (DIS) the designated processor mode. The corresponding mode status bit in the mode status register (MSTAT) is set for ENA mode and cleared for DIS mode. At reset, MSTAT is set to zero, meaning that all modes are disabled. Any number of modes can be changed in one cycle with this instruction. Multiple ENA or DIS clauses must be separated by commas.

MSTAT Bits:

0	SEC_REG	Alternate Register Data Bank
1	BIT_REV	Bit-Reverse Mode on Address Generator #1
2	AV_LATCH	ALU Overflow Status Latch Mode
3	AR_SAT	ALU AR Register Saturation Mode
4	M_MODE	MAC Result Placement Mode
5	TIMER	Timer Enable
6	G_MODE	Enables GO Mode

(Modes 4, 5, and 6 are not available for ADSP-2100 systems)

The data register bank select bit (SEC_REG) determines which set of data registers is currently active (0 = primary, 1 = secondary).

The bit-reverse mode bit (BIT_REV), when set to 1, causes addresses generated by Data Address Generator #1 to be output in bit reversed order.

The ALU overflow latch mode bit (AV_LATCH), when set to 1, causes the AV bit in the arithmetic status register to stay set once an ALU overflow occurs. In this mode, if an ALU overflow occurs, the AV bit will be set and will remain set even if subsequent ALU operations do not generate overflows. The AV bit can only be cleared by writing a zero into it directly over the DMD bus.

320

The AR saturation mode bit, (AR_SAT), when set to 1, causes the AR register to saturate if an ALU operation causes an overflow, as described in the ALU section of this document.

The MAC result placement mode (M_MODE) determines whether or not the left shift is made between the multiplier product and the MR register. (M_MODE not available on the ADSP-2100).

Setting the Timer Enable bit (TIMER) starts the timer decrementing logic. Clearing it halts the timer. (TIMER not available on the ADSP-2100).

The GO mode (G_MODE) allows an ADSP-21xx processor with internal memory to continue executing instructions (if possible) during a bus grant. The GO mode allows the processor to run; only if an external memory access is required does the processor halt, waiting for the bus to be released. (G_MODE not available on the ADSP-2100—access to external program memory is always required).

Instruction Format:
Mode Control, Instruction Type 18:

23	22	21	20	19	18	17	16	15	14	13	12	11	10	9	8	7	6	5	4	3	2	1	0
0	0	0	0	1	1	0	0	TI		MM		AS		OL		BR		SR		GM		0	0

TI:	Timer Enable	MM:	Multiplier Placement
AS:	AR Saturation Mode Control	OL:	ALU Overflow Latch Mode Control
BR:	Bit Reverse Mode Control		
GM:	GO Mode	SR:	Secondary Register Bank Mode

12 MISC
MODIFY ADDRESS REGISTER

Syntax: MODIFY   (| I0 | , | M0 |);
                          | I1 | | M1 |
                          | I2 | | M2 |
                          | I3 | | M3 |

                          | I4 | | M4 |
                          | I5 | | M5 |
                          | I6 | | M6 |
                          | I7 | | M7 |

Example: MODIFY (I1, M1);

Description: Add the selected M register (M_n) to the selected I register (I_m), then process the modified address through the modulus logic with buffer length as determined by the L register corresponding to the selected I register (L_m), and store the resulting address pointer calculation in the selected I register. The I register is modified as if an indexed memory address were taking place, but no actual memory data transfer occurs. **For linear (i.e. non-circular) indirect addressing, the L register corresponding to the I register used must be set to zero.**

The selection of the I and M registers is constrained to registers within the same Data Address Generator: selection of I0-I3 in Data Address Generator #1 constrains selection of the M registers to M0-M3. Similarly, selection of I4-I7 constrains the M registers to M4-M7.

Status Generated: None affected.

Instruction Format:
Modify Address Register, Instruction Type 21:

23	22	21	20	19	18	17	16	15	14	13	12	11	10	9	8	7	6	5	4	3	2	1	0
0	0	0	0	1	0	0	1	0	0	0	0	0	0	0	0	0	0	0	0	G	I		M

G specifies which Data Address Generator is selected. The I and M registers specified must be from the same DAG, separated by the gray bar above. I specifies the I register (depends on which DAG is selected by the G bit). M specifies the M register (depends on which DAG is selected by the G bit).

Syntax: NOP ;

Description: No operation occurs for one cycle. Execution continues with the instruction following the NOP instruction.

Status Generated: None affected.

Instruction Format:
No operation, Instruction Type 30 (see Appendix A), as shown below:

23	22	21	20	19	18	17	16	15	14	13	12	11	10	9	8	7	6	5	4	3	2	1	0
0	0	0	0	0	0	0	0	0	0	0	0	0	0	0	0	0	0	0	0	0	0	0	0

12 MULTIFUNCTION COMPUTATION with MEMORY READ

Syntax:

`<ALU>`	`, dreg =`	`DM (`	I0	,	M0	)	;
`<MAC>`			I1		M1		
`<SHIFT>`			I2		M2		
			I3		M3		
			I4		M4		
			I5		M5		
			I6		M6		
			I7		M7		
		`PM (`	I4	,	M4	)	
			I5		M5		
			I6		M6		
			I7		M7		

Permissible dregs

AX0	MX0	SI
AX1	MX1	SE
AY0	MY0	SR0
AY1	MY1	SR1
AR	MR0	
	MR1	
	MR2	

Description: Perform the designated arithmetic operation and data transfer. The read operation moves the contents of the source to the destination register. The addressing mode when combining an arithmetic operation with a memory read is register indirect with post-modify. **For linear (i.e. non-circular) indirect addressing, the L register corresponding to the I register used must be set to zero.** The contents of the source are always right-justified in the destination register.

The computation must be unconditional. All ALU, MAC and Shifter operations are permitted except Shift Immediate and ALU DIVS and DIVQ instructions.

The fundamental principle governing multifunction instructions is that registers (and memory) are read at the beginning of the processor cycle and written at the end of the cycle. The normal left-to-right order of clauses (computation first, memory read second) is intended to imply this. In fact, you may code this instruction with the order of clauses reversed. The assembler produces a warning, but the results are identical at the opcode level. If you turn off semantics checking in the assembler (using the –s switch) the warning is not issued.

324

Because of the read-first, write-second characteristic of the processor, using the same register as source in one clause and a destination in the other is legal. The register supplies the value present at the beginning of the cycle and is written with the new value at the end of the cycle.

For example,

(1) AR = AX0 + AY0, AX0 = DM (I0, M0);

is a legal version of this multifunction instruction and is not flagged by the assembler. Reversing the order of clauses, as in

(2) AX0 = DM (I0, M0) , AR = AX0 + AY0;

results in an assembler warning, but assembles and executes exactly as the first form of the instruction. Note that reading example (2) from left to right may suggest that the data memory value is loaded into AX0 and then used in the computation, all in the same cycle. In fact, this is not possible. The left-to-right logic of example (1) suggests the operation of the instruction more closely. Regardless of the apparent logic of reading the instruction from left to right, the read-first, write-second operation of the processor determines what actually happens.

Using the same register as a destination in both clauses, however, produces an indeterminate result and should not be done. The assembler issues a warning unless semantics checking is turned off. Regardless of whether or not the warning is produced, however, this practice is not supported.

The following, therefore, is illegal and not supported, even though assembler semantics checking produces only a warning:

(3) AR = AX0 + AY0, AR = DM (I0, M0);   *Illegal!*

(instruction continues on next page)

12 MULTIFUNCTION COMPUTATION with MEMORY READ

Status Generated: All status bits are affected in the same way as for the single function versions of the selected arithmetic operation.

<ALU> operation

ASTAT:	7	6	5	4	3	2	1	0
	SS	MV	AQ	AS	AC	AV	AN	AZ
	-	-	-	*	*	*	*	*

AZ	Set if result equals zero. Cleared otherwise.
AN	Set if result is negative. Cleared otherwise.
AV	Set if an overflow is generated. Cleared otherwise.
AC	Set if a carry is generated. Cleared otherwise.
AS	Affected only when executing the Absolute Value operation (ABS). Set if the source operand is negative.

<MAC> operation

ASTAT:	7	6	5	4	3	2	1	0
	SS	MV	AQ	AS	AC	AV	AN	AZ
	-	*	-	-	-	-	-	-

MV	Set if the accumulated product overflows the lower-order 32 bits of the MR register. Cleared otherwise.

<SHIFT> operation

ASTAT:	7	6	5	4	3	2	1	0
	SS	MV	AQ	AS	AC	AV	AN	AZ
	*	-	-	-	-	-	-	-

SS	Affected only when executing the EXP operation; set if the source operand is negative. Cleared if the number is positive.

Instruction Format:

ALU/MAC operation with Data Memory Read, Instruction Type 4:

23 22 21	20	19	18	17 16 15 14 13	12 11	10 9 8	7 6 5 4 3	2 1	0
0 1 1	G	0	Z	AMF	Yop	Xop	Dreg	I	M

ALU/MAC operation with Program Memory Read, Instruction Type 5:

23 22 21	20	19	18	17 16 15 14 13	12 11	10 9 8	7 6 5 4 3	2 1	0
0 1 0	1	0	Z	AMF	Yop	Xop	Dreg	I	M

Shift operation with Data Memory Read, Instruction Type 12:

23 22 21 20	19	18	17 16	15	14 13 12 11	10 9 8	7 6 5 4 3	2 1	0
0 0 0 1	0	0	1	G	0	SF	Xop	Dreg	I	M

Shift operation with Program Memory Read, Instruction Type 13:

23 22 21 20	19	18	17 16	15 14 13 12 11	10 9 8	7 6 5 4 3	2 1	0
0 0 0 1	0	0 0 1	0	SF	Xop	Dreg	I	M

Z:	Result register	Dreg:	Destination register
SF:	Shifter operation	AMF:	ALU/MAC operation
Yop:	Y operand	Xop:	X operand
G:	Data Address Generator	I:	Indirect address
M:	Modify register		register

12 MULTIFUNCTION COMPUTATION with REGISTER to REGISTER MOVE

Syntax:

| `<ALU>` |
| `<MAC>` |
| `<SHIFT>` |

`, dreg = dreg ;`

Permissible dregs

AX0	MX0	SI
AX1	MX1	SE
AY0	MY0	SR0
AY1	MY1	SR1
AR	MR0	
	MR1	
	MR2	

Description: Perform the designated arithmetic operation and data transfer. The contents of the source are always right-justified in the destination register after the read.

The computation must be unconditional. All ALU, MAC and Shifter operations are permitted except Shift Immediate and ALU DIVS and DIVQ instructions.

The fundamental principle governing multifunction instructions is that registers (and memory) are read at the beginning of the processor cycle and written at the end of the cycle. The normal left-to-right order of clauses (computation first, register transfer second) is intended to imply this. In fact, you may code this instruction with the order of clauses reversed. The assembler produces a warning, but the results are identical at the opcode level. If you turn off semantics checking in the assembler (–s switch) the warning is not issued.

Because of the read-first, write-second characteristic of the processor, using the same register as source in one clause and a destination in the other is legal. The register supplies the value present at the beginning of the cycle and is written with the new value at the end of the cycle.

For example,

(1) AR = AX0 + AY0, AX0 = MR1;

is a legal version of this multifunction instruction and is not flagged by the assembler. Reversing the order of clauses, as in

(2) AX0 = MR1, AR = AX0 + AY0;

results in an assembler warning, but assembles and executes exactly as the first form of the instruction. Note that reading example (2) from left to right may suggest that the MR1 register value is loaded into AX0 and then AX0 is used in the computation, all in the same cycle. In fact, this is not possible. The left-to-right logic of example (1) suggests the operation of the instruction more closely. Regardless of the apparent logic of reading the instruction from left to right, the read-first, write-second operation of the processor determines what actually happens.

Using the same register as a destination in both clauses, however, produces an indeterminate result and should not be done. The assembler issues a warning unless semantics checking is turned off. Regardless of whether or not the warning is produced, however, this practice is not supported.

The following, therefore, is illegal and not supported, even though assembler semantics checking produces only a warning:

(3) AR = AX0 + AY0, AR = MR1;   *Illegal!*

Status Generated: All status bits are affected in the same way as for the single function versions of the selected arithmetic operation.

<ALU> operation

ASTAT:	7	6	5	4	3	2	1	0
	SS	MV	AQ	AS	AC	AV	AN	AZ
	-	-	-	*	*	*	*	*

AZ	Set if result equals zero. Cleared otherwise.
AN	Set if result is negative. Cleared otherwise.
AV	Set if an overflow is generated. Cleared otherwise.
AC	Set if a carry is generated. Cleared otherwise.
AS	Affected only when executing the Absolute Value operation (ABS). Set if the source operand is negative.

(instruction continues on next page)

12 MULTIFUNCTION
COMPUTATION with REGISTER to REGISTER MOVE

<MAC> operation

ASTAT:

7	6	5	4	3	2	1	0
SS	MV	AQ	AS	AC	AV	AN	AZ
-	*	-	-	-	-	-	-

MV   Set if the accumulated product overflows the lower-order 32
            bits of the MR register. Cleared otherwise.

<SHIFT> operation

ASTAT:

7	6	5	4	3	2	1	0
SS	MV	AQ	AS	AC	AV	AN	AZ
*	-	-	-	-	-	-	-

SS   Affected only when executing the EXP operation; set if the
            source operand is negative. Cleared if the number is
            positive.

Instruction Format:
ALU/MAC operation with Data Register Move, Instruction Type 8:

23	22	21	20	19	18	17 16 15 14 13 12	11 10 9	8 7 6	5 4 3	2 1 0
0	0	1	0	1	Z	AMF	Yop	Xop	Dreg dest	Dreg source

Shift operation with Data Register Move, Instruction Type 14:

23	22	21	20	19	18	17 16 15	14 13 12 11	10 9	8 7 6	5 4 3	2 1 0
0	0	0	1	0	0	0 0 0	SF	Xop	Dreg dest	Dreg source	

Z:	Result register	Dreg:	Data register
SF:	Shifter operation	AMF:	ALU/MAC operation
Yop:	Y operand	Xop:	X operand

Syntax:

DM (
| I0 | , | M0 |)
|----|---|----|
| I1 | | M1 |
| I2 | | M2 |
| I3 | | M3 |

I4		M4
I5		M5
I6		M6
I7		M7

= dreg ,
\<ALU\>	;
\<MAC\>	
\<SHIFT\>	

PM (
| I4 | , | M4 |)
|----|---|----|
| I5 | | M5 |
| I6 | | M6 |
| I7 | | M7 |

Permissible dregs

AX0	MX0	SI
AX1	MX1	SE
AY0	MY0	SR0
AY1	MY1	SR1
AR	MR0	
	MR1	
	MR2	

Description: Perform the designated arithmetic operation and data transfer. The write operation moves the contents of the source to the specified memory location. The addressing mode when combining an arithmetic operation with a memory write is register indirect with post-modify. **For linear (i.e. non-circular) indirect addressing, the L register corresponding to the I register used must be set to zero.** The contents of the source are always right-justified in the destination register.

The computation must be unconditional. All ALU, MAC and Shifter operations are permitted except Shift Immediate and ALU DIVS and DIVQ instructions.

The fundamental principle governing multifunction instructions is that registers (and memory) are read at the beginning of the processor cycle and written at the end of the cycle. The normal left-to-right order of clauses (memory write first, computation second) is intended to imply this. In fact, you may code this instruction with the order of clauses reversed. The assembler produces a warning, but the results are identical at the opcode level. If you turn off semantics checking in the assembler (–s switch) the warning is not issued.

(instruction continues on next page)   **331**

12 MULTIFUNCTION COMPUTATION with MEMORY WRITE

Because of the read-first, write-second characteristic of the processor, using the same register as destination in one clause and a source in the other is legal. The register supplies the value present at the beginning of the cycle and is written with the new value at the end of the cycle.

For example,

(1) DM (I0, M0) = AR, AR = AX0 + AY0;

is a legal version of this multifunction instruction and is not flagged by the assembler. Reversing the order of clauses, as in

(2) AR = AX0 + AY0, DM (I0, M0) = AR;

results in an assembler warning, but assembles and executes exactly as the first form of the instruction. Note that reading example (2) from left to right may suggest that the result of the computation in AR is then written to memory, all in the same cycle. In fact, this is not possible. The left-to-right logic of example (1) suggests the operation of the instruction more closely. Regardless of the apparent logic of reading the instruction from left to right, the read-first, write-second operation of the processor determines what actually happens.

Status Generated: All status bits are affected in the same way as for the single function versions of the selected arithmetic operation.

<ALU> operation

ASTAT:	7	6	5	4	3	2	1	0
	SS	MV	AQ	AS	AC	AV	AN	AZ
	-	-	-	*	*	*	*	*

AZ	Set if result equals zero. Cleared otherwise.
AN	Set if result is negative. Cleared otherwise.
AV	Set if an overflow is generated. Cleared otherwise.
AC	Set if a carry is generated. Cleared otherwise.
AS	Affected only when executing the Absolute Value operation (ABS). Set if the source operand is negative.

332

\<MAC\> operation

ASTAT:	7	6	5	4	3	2	1	0
	SS	MV	AQ	AS	AC	AV	AN	AZ
	-	*	-	-	-	-	-	-

MV   Set if the accumulated product overflows the lower-order 32 bits of the MR register. Cleared otherwise.

\<SHIFT\> operation

ASTAT:	7	6	5	4	3	2	1	0
	SS	MV	AQ	AS	AC	AV	AN	AZ
	*	-	-	-	-	-	-	-

SS   Affected only when executing the EXP operation; set if the source operand is negative. Cleared if the number is positive.

Instruction Format:
ALU/MAC operation with Data Memory Write, Instruction Type 4:

23	22	21	20	19	18	17 16 15 14 13	12 11	10 9 8	7 6 5 4	3 2	1 0
0	1	1	G	1	Z	AMF	Yop	Xop	Dreg	I	M

ALU/MAC operation with Program Memory Write, Instruction Type 5:

23	22	21	20	19	18	17 16 15 14 13	12 11	10 9 8	7 6 5 4	3 2	1 0
0	1	0	1	1	Z	AMF	Yop	Xop	Dreg	I	M

(instruction continues on next page)

333

12 MULTIFUNCTION COMPUTATION with MEMORY WRITE

Shift operation with Data Memory Write, Instruction Type 12:

23	22	21	20	19	18	17	16	15	14	13	12	11	10 9 8	7 6 5	4 3 2	1	0
0	0	0	1	0	0	1	G	1				SF	Xop	Dreg	I	M

Shift operation with Program Memory Write, Instruction Type 13:

23	22	21	20	19	18	17	16	15	14	13	12	11	10 9 8	7 6 5	4 3 2	1	0
0	0	0	1	0	0	0	1	1				SF	Xop	Dreg	I	M

Z:	Result register	Dreg:	Destination register
SF:	Shifter operation	AMF:	ALU/MAC operation
Yop:	Y operand	Xop:	X operand
I:	Indirect address register	M:	Modify register
G:	Data Address Generator; I & M registers must be from the same DAG, as separated by the gray bar in the Syntax description.		

Syntax:

AX0	= DM (	I0	,	M0	) ,	AY0	= PM (	I4	,	M4	) ;
AX1		I1		M1		AY1		I5		M5	
MX0		I2		M2		MY0		I6		M6	
MX1		I3		M3		MY1		I7		M7	

Description: Perform the designated memory reads, one from data memory and one from program memory. Each read operation moves the contents of the memory location to the destination register. For this double data fetch, the destinations for data memory reads are the X registers in the ALU and the MAC, and the destinations for program memory reads are the Y registers. The addressing mode for this memory read is register indirect with post-modify. **For linear (i.e. non-circular) indirect addressing, the L register corresponding to the I register used must be set to zero.** The contents of the source are always right-justified in the destination register.

For information on extra cycle conditions, refer to the Instruction Set Overview at the beginning of this chapter.

Status Generated: No status bits are affected.

Instruction Format:
ALU/MAC with Data & Program Memory Read, Instruction Type 1:

23 22	21 20	19 18	17 16 15 14 13	12 11 10 9 8	7	6	5	4 3 2 1 0
1 1	PD	DD	AMF	0 0 0 0 0	PM I	PM M	DM I	DM M

AMF specifies the ALU or MAC function. In this case, AMF = 00000, designating a no-operation for the ALU or MAC function.

PD:	Program Destination register	DD:	Data Destination register
AMF:	ALU/MAC operation	I:	Indirect address register
M:	Modify register		

12 MULTIFUNCTION
ALU / MAC with DATA & PROGRAM MEMORY READ

Syntax:

$$
\begin{vmatrix} \text{<ALU>} \\ \text{<MAC>} \end{vmatrix} , \begin{vmatrix} \text{AX0} \\ \text{AX1} \\ \text{MX0} \\ \text{MX1} \end{vmatrix} = \text{DM} (\begin{vmatrix} \text{I0} \\ \text{I1} \\ \text{I2} \\ \text{I3} \end{vmatrix} , \begin{vmatrix} \text{M0} \\ \text{M1} \\ \text{M2} \\ \text{M3} \end{vmatrix}), \begin{vmatrix} \text{AY0} \\ \text{AY1} \\ \text{MY0} \\ \text{MY1} \end{vmatrix} = \text{PM} (\begin{vmatrix} \text{I4} \\ \text{I5} \\ \text{I6} \\ \text{I7} \end{vmatrix} , \begin{vmatrix} \text{M4} \\ \text{M5} \\ \text{M6} \\ \text{M7} \end{vmatrix});
$$

Description: This instruction combines an ALU or a MAC operation with a data memory read and a program memory read. The read operations move the contents of the memory location to the destination register. For this double data fetch, the destinations for data memory reads are the X registers in the ALU and the MAC, and the destinations for program memory reads are the Y registers. The addressing mode is register indirect with post-modify. **For linear (i.e. non-circular) indirect addressing, the L register corresponding to the I register used must be set to zero.** The contents of the source are always right-justified in the destination register after the read.

The computation must be unconditional. All ALU and MAC operations are permitted except the DIVS and DIVQ instructions. The results of the computation must be written into the R register of the computational unit; ALU results to AR, MAC results to MR.

The fundamental principle governing multifunction instructions is that registers (and memory) are read at the beginning of the processor cycle and written at the end of the cycle. The normal left-to-right order of clauses (computation first, memory reads second) is intended to imply this. In fact, you may code this instruction with the order of clauses altered. The assembler produces a warning, but the results are identical at the opcode level. If you turn off semantics checking in the assembler (–s switch) the warning is not issued.

The same data register may be used as a source for the arithmetic operation and as a destination for the memory read. The register supplies the value present at the beginning of the cycle and is written with the value from memory at the end of the cycle.

For example,

(1) MR=MR+MX0*MY0(UU), MX0=DM(I0, M0), MY0=PM(I4,M4);

is a legal version of this multifunction instruction and is not flagged by the assembler. Changing the order of clauses, as in

(2) MX0=DM(I0, M0), MY0=PM(I4,M4), MR=MR+MX0*MY0(UU);

results in an assembler warning, but assembles and executes exactly as the first form of the instruction. Note that reading example (2) from left to right may suggest that the data memory value is loaded into MX0 and MY0 and subsequently used in the computation, all in the same cycle. In fact, this is not possible. The left-to-right logic of example (1) suggests the operation of the instruction more closely. Regardless of the apparent logic of reading the instruction from left to right, the read-first, write-second operation of the processor determines what actually happens.

Status Generated: All status bits are affected in the same way as for the single operation version of the selected arithmetic operation.

<ALU> operation

ASTAT:	7	6	5	4	3	2	1	0
	SS	MV	AQ	AS	AC	AV	AN	AZ
	-	-	-	*	*	*	*	*

AZ   Set if result equals zero. Cleared otherwise.
AN   Set if result is negative. Cleared otherwise.
AV   Set if an overflow is generated. Cleared otherwise.
AC   Set if a carry is generated. Cleared otherwise.
AS   Affected only when executing the Absolute Value operation (ABS). Set if the source operand is negative.

<MAC> operation

ASTAT:	7	6	5	4	3	2	1	0
	SS	MV	AQ	AS	AC	AV	AN	AZ
	-	*	-	-	-	-	-	-

MV   Set if the accumulated product overflows the lower-order 32-bits of the MR register. Cleared otherwise.

(instruction continues on next page)

12 MULTIFUNCTION
ALU / MAC with DATA & PROGRAM MEMORY READ

Instruction Format:

ALU/MAC with Data and Program Memory Read, Instruction Type 1:

23 22	21 20	19 18	17 16 15 14 13	12 11	10 9 8	7 6 5 4 3 2 1 0

23	22	21 20	19 18	17 16 15 14 13	12 11	10 9 8	7 6 5	4 3	2 1	0
1	1	PD	DD	AMF	Yop	Xop	PM I	PM M	DM I	DM M

PD: Program Destination register   DD: Data Destination register
AMF: ALU/MAC operation   M: Modify register
Yop: Y operand   Xop: X operand
I: Indirect address register

338

Hardware Examples ■ 13

13.1   OVERVIEW

This chapter describes some hardware examples of additional circuits that can be added to the serial ports, the host interface port (HIP) or the memory port. These examples are in addition to the normal serial port interface, HIP interface and memory interface information found in Chapters 5, 7, and 10 respectively. As with any hardware design, it is important that timing information be carefully analyzed. Therefore, the data sheet for the particular ADSP-2100 family processor used should be referenced to augment the material found in this chapter.

13.2   BOOT LOADING FROM HOST USING BUS REQUEST/BUS GRANT

All ADSP-2100 family processors, that have internal program RAM, support boot loading. With boot loading, the processor reads instructions from a byte-wide external memory device (usually an EPROM) over the memory interface and stores the instructions in the 24-bit wide internal program memory. Once the external memory device is set up to provide bytes in the proper order, the boot operation can run automatically and transparently at reset or when forced in software. See Chapter 10, "Memory Interface."

In some systems where the ADSP-2100 family processor is controlled by a host, it is necessary to boot the ADSP-2100 family processor directly from the host processor. In this case the host, rather than an EPROM, is the source of bytes to be loaded into internal memory. If the ADSP-2100 family processor has a host interface port, it can perform boot loading automatically through this port. If the processor does not have a host interface port, however, it can still boot through the memory interface using bus request and the circuitry described in this section.

13 Hardware Examples

13.2.1 Boot Loading Circuit Design

The schematic of Figure 13.1 shows the basic circuit for boot loading. This design uses the Bus Request ($\overline{BR}$) and Bus Grant ($\overline{BG}$) pins of the ADSP-21xx to control the transfer of data from the host to the ADSP-21xx. This requires the ADSP-21xx to run with a minimum of 1 boot memory wait state to assure that the $\overline{BR}$ is recognized every boot memory read cycle.

In this design the ADSP-21xx is memory mapped into two locations of the host's memory space, one location for data and one for control signals. It assumes that the proper host memory address decoding circuitry exists and supplies the three control signals, Host Load 0 (HL), Host Load 1 (HL1) and Host Read 0 (HR0). This design uses nine host data bits (D0-D8), where D0–D7 are used for data, D0–D2 are also used for controlling the booting and D8 is used for reading the ADSP-21xx's $\overline{BG}$ signal.

The control signal HL0 is used to latch the boot data into the register (U1). HL1 is used to latch the control bits D0–D2 into the registers (U2 and U3A) and HR0 is used to read the $\overline{BG}$ signal of the ADSP-21xx through buffer (U4).

The host initiates a boot by writing the first byte of boot data into the data register U1. Then the host must set the control bits by writing to the control registers (U2 and U3A). D0 is used to reset the ADSP-21xx, D1 is used to request the ADSP-21xx bus and D2 is used to clear the host bus request.

After the first byte is written into the data register, the host must clear the control bits D0–D2 to 0. This places the ADSP-21xx into reset (D0), enables the host bus request (D1), sets the host bus request clear (D2).

The booting is started when the ADSP-21xx is brought out of reset by the host writing a 1 to D0. When MMAP=0, the first thing the ADSP-21xx does when it comes out of reset is assert the boot memory select ($\overline{BMS}$) signal and begin to read the first location of boot memory (usually a PROM), but in this case the data is read from register, U1. In this circuit when $\overline{BMS}$ goes low, the preset input of flip-flop U3B is brought low, requesting the ADSP-21xx bus via the bus request ($\overline{BR}$) signal. When the ADSP-21xx bus is requested, the processor finishes the current instruction, in this case the read of the first boot data byte, and suspends operation putting its buses and control signals in a high-impedance state. The processor also asserts its bus grant ($\overline{BG}$) output signal indicating that the bus has been relinquished.

Hardware Examples 13

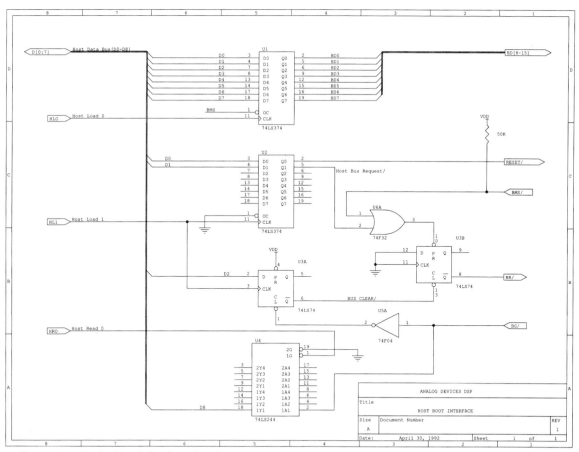

Figure 13.1 Basic Circuit For Boot Loading

13 Hardware Examples

While the ADSP-21xx processor is suspended, the host writes the next byte of boot data into U1, and then clears the $\overline{BR}$ signal by writing a 1 to the control bit D2. When $\overline{BR}$ is cleared the ADSP-21xx resumes the boot process from where it had stopped. The $\overline{BMS}$ signal goes low which causes the $\overline{BR}$ to be asserted once again. The ADSP-21xx will finish the current read and grant the bus. This process is repeated until all the boot bytes are loaded. When the last byte is read by ADSP-21xx, the host bus request has to be disabled by writing a 1 to control bit D1 before the $\overline{BR}$ is cleared. This will deassert the host $\overline{BR}$ and allow the ADSP-21xx to begin execution of the downloaded code.

A flow chart of the boot loading process is shown in Figure 13.2. Figure 13.3 shows typical timing.

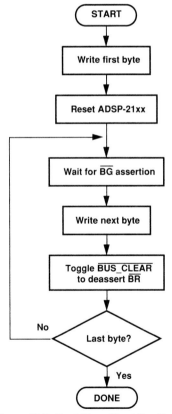

Figure 13.2 Host Flow Chart For Boot Loading

Hardware Examples 13

The host must provide bytes in the sequence expected by the ADSP-2100 family processor. This sequence is described in the Chapter 10, "Memory Interface."

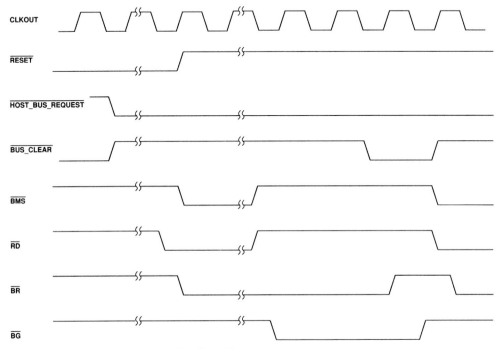

Figure 13.3 Timing For Boot Loading From Host

13.3   SERIAL PORT TO CODEC INTERFACE

A codec (COder/DECoder) incorporates analog-to-digital conversion, digital-to-analog conversion, and filtering in one device. The codec shown in this example also performs pulse-code modulation (PCM) encoding and decoding according to the CCITT μ-law standard. PCM compresses digital data so that fewer bits are needed to store the same information. The ADSP-21xx serial ports have both μ-law and A-law companding (compressing/expanding) capability.

13 Hardware Examples

In the example described here, a codec converts its analog input to digital data, compresses it and sends it serially to the SPORT on an ADSP-21xx processor. At the same time, the processor sends compressed serial data via the SPORT to the codec, which expands the data and converts the result to an analog signal.

Figure 13.4 shows an industry standard μ-law companding codec connected to a serial port (in this case, SPORT0) on an ADSP-21xx processor. The codec's analog input at VFXI+ is internally amplified by a gain which is controlled by the resistor combination at GSX and VFXI–. The gain is

$$20 \times \log (R1 + R2)/R2$$

in this case, 20 log 2.

The ADSP-21xx controls codec operation by supplying master and bit clock signals. In the configuration shown, the codec transmit and receive sections operate synchronously. MCLKR and MCLKX are the master clocks for the receive and transmit sections of the codec, respectively. BCLKX is the bit clock and in this configuration is used for clocking both received and transmitted serial data. MCLKR, MCLKX and BCLKX must be synchronous and in this case they are the same signal, namely the SCLK0 output generated by the ADSP-21xx processor. The BCLKR/CLKSEL input, tied low, selects the frequency of MCLKX to be 2.048 MHz. The ADSP-21xx must be programmed for internal SCLK0 generation at 2.048 MHz.

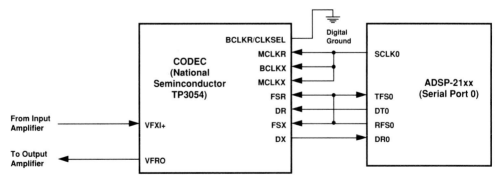

Figure 13.4 ADSP-21xx Serial Port (SPORT0) To CODEC

Hardware Examples 13

The processor uses frame synchronization signals to tell the codec to send and receive data. To transmit data to the codec, it sends a TFS0 pulse to the FSR input of the codec and then outputs the eight bits on DT0 on the next eight serial clock periods. The codec receives the data on its DR input. Likewise, the processor initiates a data receive operation by sending an RFS0 pulse to the codec's FSX input, which causes the codec to output eight bits on its DX output on the next eight serial clock periods. The processor receives the data on its DR0 input. The ADSP-21xx must be programmed to use normal framing, 8-bit data words, and internal, active-high frame sync generation.

The ADSP-21xx code shown in Listing 13.1 configures SPORT0 for operation as required in this example:

- Internally generated serial clock
- 2.048 MHz serial clock frequency
- Both transmit and receive frame syncs required
- Use normal framing for both transmit and receive
- Internally generated transmit and receive frame syncs
- Both frame syncs active high
- Word length of eight bits
- μ-law companding

This code assumes the processor operating at 12.288 MHz. The code also sets up the processor to request data from the codec at an 8 kHz rate (this register is not initialized at reset and should always be written before the SPORT is enabled if RFS is generated internally). The processor transmits data as needed by the program it is executing.

```
AX0=0x6927;        {Int SCLK, RFS/TFS req, norm framing,}
DM(0x3FF6)=AX0;    {generate RFS, active HI, Mu-law, word length 8}

AX0=2;             {value of SCLKDIV for 2.048 MHz}
DM(0x3FF5)=AX0;    {with a 12.888 MHz CLKOUT}

AX0=255;           {RFSDIV=256, 256 SCLKs between}
DM(0x3FF4)=AX0;    {frame syncs, 8 kHz framing}

AX0=0x1038;        {enable SPORT0 only, leave defaults}
DM(0x3FFF)=AX0;
```

Listing 13.1 Serial Port Initialization Example

13 Hardware Examples

13.4   SERIAL PORT TO DAC INTERFACE

Any DSP process must ultimately output analog information. The serial port of the ADSP-21xx processors can send data directly to a DAC (digital-to-analog converter) for conversion to an analog signal.

Analog Devices' AD766 is one DAC that requires no extra logic to interface to the SPORT. The AD766 receives 16-bit data words serially, MSB first, which it then converts to an analog signal. Its digital interface consists of three inputs: DATA, the serial data input; $\overline{CLK}$, for clocking data into the DAC (active low because data is clocked on the falling edge) and LE (latch enable), which latches each 16-bit word into the conversion section of the DAC.

The serial port connection to the AD766 is shown in Figure 13.5. In this configuration, the processor generates SCLK internally and provides it to the DAC. Serial data is output from the DT pin to the DATA input of the DAC. The TFS signal provides the DAC's LE input.

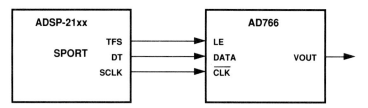

Figure 13.5 Serial Port Interface To AD766 DAC

LE should go low on the clock cycle after the LSB (sixteenth bit) of a word is transmitted, to latch the 16-bit word into the DAC. To provide this timing, TFS is configured for the alternate framing mode, non-inverted; it goes high when the first bit is transmitted and low after the last bit is transmitted. This low-going edge latches the word into the AD766. The only restriction is that the SPORT cannot transmit continuously; there must be a break in between the last bit of one word and the first bit of the next so that TFS can go low. Figure 13.6 shows the timing.

Hardware Examples 13

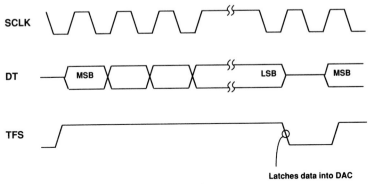

Figure 13.6 SPORT To AD766 DAC Timing

The configuration of the SPORT control register for this application is shown in Figure 13.7.

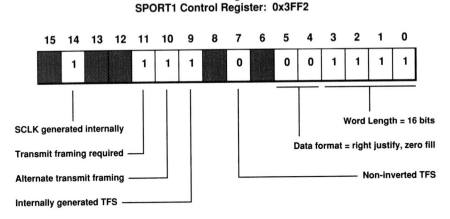

Figure 13.7 SPORT To AD766 DAC Control Register Settings

13 Hardware Examples

13.5   SERIAL PORT TO ADC INTERFACE

An ADC (analog-to-digital converter) converts an analog signal to digital samples that a DSP processor can operate on. The ADSP-21xx processors can receive data from an ADC directly through a serial port.

Analog Devices' AD7872 is one ADC that requires no extra logic to interface to the SPORT. The AD7872 converts an analog signal to 14-bit samples. Each sample is padded with two zero MSBs to yield 16-bit samples. The AD7872 outputs each sample serially, MSB first. Its digital interface consists of three pins: SDATA, the serial data output; SCLK, for clocking data out; and $\overline{\text{SSTRB}}$, (serial strobe), which frames each serial word.

The serial port connection to the AD7872 is shown in Figure 13.8. The timer regulates sampling via the $\overline{\text{CONVST}}$ input at a constant frequency. Instead of the timer, an unused serial clock or flag output from the ADSP-21xx processor can be programmed to generate the $\overline{\text{CONVST}}$ signal. The AD7872 generates SCLK internally and provides it to the processor. With the CONTROL input held at –5 V, the SCLK signal is continuous, running even when no data is being output.

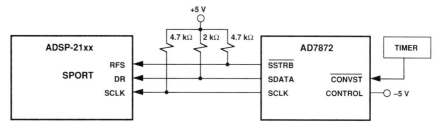

Figure 13.8 Serial Port Interface To AD7872 ADC

Serial data is output from the SDATA output of the ADC to the processor's DR pin. The $\overline{\text{SSTRB}}$ signal provides the RFS input to the processor. $\overline{\text{SSTRB}}$ goes low when the first bit is transmitted to the processor. Figure 13.9 shows the timing of the serial data transfer.

Hardware Examples 13

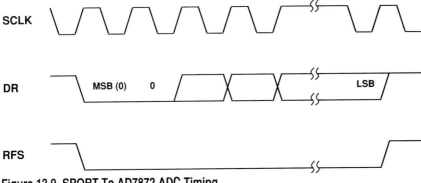

Figure 13.9 SPORT To AD7872 ADC Timing

RFS is configured for the alternate framing mode, externally generated, with inverted (active low) logic. The SPORT must also be programmed for external serial clock and a serial word length of 16 bits. The configuration of the SPORT control register for this application is shown in Figure 13.10.

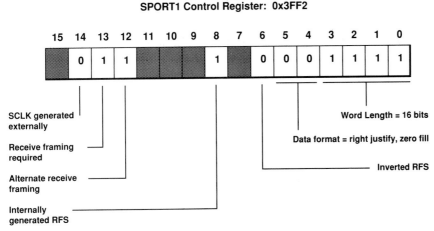

Figure 13.10 SPORT To AD7872 ADC Control Register Settings

13 Hardware Examples

13.6 SERIAL PORT TO SERIAL PORT INTERFACE

The serial ports provide a convenient way to transfer data between ADSP-21xx processors without using external memory or the memory bus and without halting either processor. The serial ports are connected as shown in Figure 13.11; in this example, SPORT1 of processor #1 is connected to SPORT0 of processor #2.

The serial clock used by both processors is generated internally by processor #1. Processor #2 is configured to receive its serial clock externally. The serial port control registers should be set up with the following parameters.

Processor 1, SPORT1	Processor 2, SPORT0
SCLKDIV = system-dependent	SCLKDIV = system-dependent
SLEN = system-dependent	SLEN = system-dependent
ISCLK = 1	ISCLK = 0
TFSR = 1	TFSR = 1
RFSR = 1	RFSR = 1
IRFS = 0	IRFS = 0
ITFS = 1	ITFS = 1
RFSDIV = don't care	RFSDIV = don't care

TFSW1 = RFSW1 = TFSW2 = RFSW2 = system-dependent
INVRFS1 = INVTFS1 = INVRFS2 = INVTFS2 = system-dependent

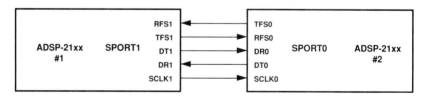

Figure 13.11 Serial Port Interface Between Two ADSP-21xx Processors

Frame synchronization is used to coordinate the transfer of serial data. Each processor generates a transmit frame sync (TFS) signal internally and expects to receive its receive frame sync (RFS) signal externally, from the other processor. The framing mode can be normal or alternate, but must be the same for both SPORTs. Likewise, the SPORTs must be configured for the same serial word length and companding type, if companding is used, or data format if companding is not used.

Hardware Examples 13

The autobuffering capability of the serial ports can be used in this configuration to transfer an entire buffer of data from the data memory space of one processor to the other's, *without interrupt overhead*. The serial ports handshake automatically—when one processor writes its' TX0 register, the data is automatically transmitted to the other processor's RX0 register and an autobuffer cycle is generated.

In fact, autobuffer transfers can occur in both directions at the same time, in the background, while each processor is executing some other primary function. Each SPORT will generate an interrupt when the autobuffer transfer is complete. The description of autobuffering in the Serial Port chapter shows an example of the code for setting up autobuffering.

13.7   80C51 INTERFACE TO HOST INTERFACE PORT

The host interface port (HIP) on the ADSP-2111 and ADSP-21msp50 processors facilitates communication with a host microcomputer such as the Intel 80C51. An example connection is shown in Figure 13.12. In this example, the HIP data registers (HDRs) and HIP status registers (HSRs) of the ADSP-21xx processor occupy eight contiguous locations in the memory space of the 80C51.

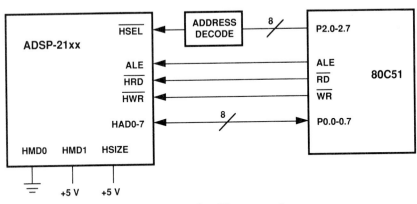

Figure 13.12 Host Port Interface To 80C51 Microcomputer

To access one of the HIP registers, the 80C51 asserts ALE and outputs a 16-bit address, with the upper half on P2.0-2.7 and the lower half on P0.0-0.7. The upper half is decoded to select the HIP via HSEL, and the lower

13 Hardware Examples

half selects the HIP register via HAD0-7. The ALE assertion causes the HIP to latch the address so that the 8-bit data can then be transferred on the HAD0-7 lines. The 80C51 asserts $\overline{WR}$ for a write or $\overline{RD}$ for a read.

In this example, the 80C51 reads and writes 8-bit data, so the ADSP-21xx processor's HSIZE input is tied high. Only the lower eight bits of each HIP register are used. HMD0 is tied low because the 80C51 uses separate read and write strobes rather than a single Read/$\overline{Write}$ line. HMD1 is tied high because the address and data use the same bus (time-multiplexed using ALE) rather than separate buses.

Software Examples ◼ 14

14.1 OVERVIEW

This chapter provides a brief summary of the development process that you use to create executable programs for the ADSP-2100 family. The summary is followed by a number of software examples that can give you an idea of how to write your own applications.

The software examples presented in this chapter are used a variety of DSP operations. The FIR filter and cascaded biquad IIR filter are general filter algorithms that can be tailored to many applications. Matrix multiplication is used in image processing and other areas requiring vector operations. The sine function is required for many scientific calculations. The FFT (fast Fourier transform) has wide application in signal analysis. Each of these examples is described in greater detail in *Digital Signal Processing Applications Using The ADSP-2100 Family Volume 1* available from Prentice Hall. They are presented here to show some aspects of typical programs.

The FFT example is a complete program, showing a subroutine that performs the FFT and a main calling program that initializes registers and calls the FFT subroutine as well as an auxiliary routine.

Each of the other examples is shown as a subroutine in its own module. The module starts with a .MODULE directive that names the module and ends with the .ENDMOD directive. The subroutine can be called from a program in another module that declares the starting label of the subroutine as an external symbol. This is the same label that is declared with the .ENTRY directive in the subroutine module. The last instruction in each subroutine is the RTS instruction, which returns control to the calling program. Notice that all comments are within curly brackets { }.

14 Software Examples

Each module is prefaced by a comment block that provides the following information:

Calling Parameters	Register values that the calling program must set before calling the subroutine
Return Values	Registers that hold the results of the subroutine
Altered Registers	Registers used by the subroutine. The calling program must save them before calling the subroutine and restore them afterward if it needs to preserve their values.
Computation Time	The number of instruction cycles needed to perform the subroutine

14.2   SYSTEM DEVELOPMENT PROCESS

The ADSP-2100 family of processors is supported by a complete set of development tools. Programming aids and processor simulators facilitate software design and debug. In-circuit emulators and demonstration systems help in hardware prototyping.

The software development system includes several programs: System Builder, Assembler, Linker, PROM Splitter, Simulators and C Compiler with Runtime Library. These programs are described in detail in the *ADSP-2100 Family Assembler Manual* and the *ADSP-2100 Family Simulator Manual*.

Figure 14.1 shows a flow chart of the ADSP-2100 family system development process.

The development process begins with the task of describing the chosen hardware environment to the development software. You create a system specification file using a text editor. This file contains simple directives that describe the locations of memory and I/O ports, the type of processor, and the state of the MMAP pin in the target hardware configuration. The system builder reads this file and generates an architecture description file which passes information to the linker, simulator and emulator.

Software Examples 14

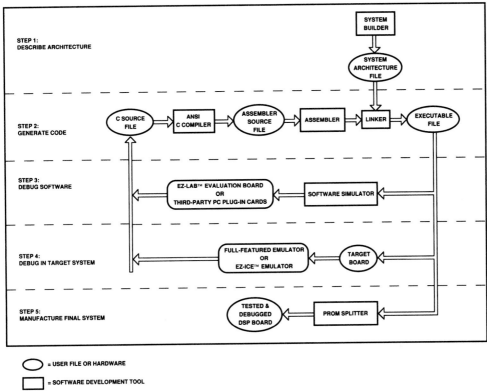

STEP 1:
DESCRIBE ARCHITECTURE

SYSTEM BUILDER

SYSTEM ARCHITECTURE FILE

STEP 2:
GENERATE CODE

C SOURCE FILE

ANSI C COMPILER

ASSEMBLER SOURCE FILE

ASSEMBLER

LINKER

EXECUTABLE FILE

STEP 3:
DEBUG SOFTWARE

EZ-LAB™ EVALUATION BOARD OR THIRD-PARTY PC PLUG-IN CARDS

SOFTWARE SIMULATOR

STEP 4:
DEBUG IN TARGET SYSTEM

FULL-FEATURED EMULATOR OR EZ-ICE™ EMULATOR

TARGET BOARD

STEP 5:
MANUFACTURE FINAL SYSTEM

TESTED & DEBUGGED DSP BOARD

PROM SPLITTER

◯ = USER FILE OR HARDWARE

▢ = SOFTWARE DEVELOPMENT TOOL

◯ = HARDWARE DEVELOPMENT TOOL

Figure 14.1 ADSP-2100 Family System Development Process

You begin code generation by creating source code files in C language or assembly language. A module is a unit of assembly language comprising a main program, subroutine, or data variable declarations. C programmers write C language files and use the C compiler to create assembly code modules from them. Assembly language programmers write assembly code modules directly. Each code module is assembled separately by the assembler.

The linker links several modules together to form an executable program (memory image file). The linker reads the target hardware information from the architecture description file to determine appropriate addresses for code and data. In the assembly modules you may specify each code/data fragment as completely relocatable, relocatable within a defined

355

14 Software Examples

memory segment, or non-relocatable (placed at an absolute address). The linker places non-relocatable code or data modules at the specified memory addresses, provided the memory area has the correct attributes. Relocatable objects are placed at addresses selected by the linker. The linker generates a memory image file containing a single executable program which may be loaded into a simulator or emulator for testing.

The simulator provides windows that display different portions of the hardware environment. To replicate the target hardware, the simulator configures its memory according to the architecture description file generated by the system builder, and simulates memory-mapped I/O ports. This simulation allows you to debug the system and analyze performance before committing to a hardware prototype.

After fully simulating your system and software, you can use an emulator in the prototype hardware to test circuitry, timing, and real-time software execution. The emulator has overlay memory which can be used in place of target system memory components.

The PROM splitter software tool translates the linker-output program (memory image file) into an industry-standard file format for a PROM programmer. Once you program the code in PROM devices and install an ADSP-2100 family processor into your prototype, it is ready to run.

14.3   SINGLE-PRECISION FIR TRANSVERSAL FILTER

An FIR transversal filter structure can be obtained directly from the equation for discrete-time convolution.

$$y(n) = \sum_{k=0}^{N-1} h_k(n) \, x(n-k)$$

In this equation, $x(n)$ and $y(n)$ represent the input to and output from the filter at time n. The output $y(n)$ is formed as a weighted linear combination of the current and past input values of x, $x(n-k)$. The weights, $h_k(n)$, are the transversal filter coefficients at time n. In the equation, $x(n-k)$ represents the past value of the input signal "contained" in the $(k+1)$th tap of the transversal filter. For example, $x(n)$, the present value of the input signal, would correspond to the first tap, while $x(n-42)$ would correspond to the forty-third filter tap.

Software Examples 14

The subroutine that realizes the sum-of-products operation used in computing the transversal filter is shown in Listing 14.1.

```
.MODULE fir_sub;

{
    FIR Transversal Filter Subroutine

    Calling Parameters
        I0 -> Oldest input data value in delay line
        L0 = Filter length (N)
        I4 -> Beginning of filter coefficient table
        L4 = Filter length (N)
        M1,M5 = 1
        CNTR = Filter length - 1 (N-1)

    Return Values
        MR1 = Sum of products (rounded and saturated)
        I0 -> Oldest input data value in delay line
        I4 -> Beginning of filter coefficient table

    Altered Registers
        MX0,MY0,MR

    Computation Time
        N - 1 + 5 + 2 cycles

    All coefficients and data values are assumed to be
    in 1.15 format.
}

.ENTRY  fir;

fir:    MR=0, MX0=DM(I0,M1), MY0=PM(I4,M5);
        DO sop UNTIL CE;
sop:        MR=MR+MX0*MY0(SS), MX0=DM(I0,M1), MY0=PM(I4,M5);
        MR=MR+MX0*MY0(RND);
        IF MV SAT MR;
        RTS;
.ENDMOD;
```

Listing 14.1 Single-Precision FIR Transversal Filter

14 Software Examples

14.4 CASCADED BIQUAD IIR FILTER

A second-order biquad IIR filter section is represented by the transfer function (in the z-domain):

$$H(z) = Y(z)/X(z) = (B_0 + B_1 z^{-1} + B_2 z^{-2})/(1 + A_1 z^{-1} + A_2 z^{-2})$$

where A_1, A_2, B_0, B_1 and B_2 are coefficients that determine the desired impulse response of the system H(z). The corresponding difference equation for a biquad section is:

$$Y(n) = B_0 X(n) + B_1 X(n-1) + B_2 X(n-2) - A_1 Y(n-1) - A_2 Y(n-2)$$

Higher-order filters can be obtained by cascading several biquad sections with appropriate coefficients. The biquad sections can be scaled separately and then cascaded in order to minimize the coefficient quantization and the recursive accumulation errors.

A subroutine that implements a high-order filter is shown in Listing 14.2. A circular buffer in program memory contains the scaled biquad coefficients. These coefficients are stored in the order: B_2, B_1, B_0, A_2 and A_1 for each biquad. The individual biquad coefficient groups must be stored in the order that the biquads are cascaded.

```
.MODULE      biquad_sub;

{       Nth order cascaded biquad filter subroutine

        Calling Parameters:

          SR1=input X(n)
          I0 -> delay line buffer for X(n-2), X(n-1),
             Y(n-2), Y(n-1)
          L0 = 0
          I1 -> scaling factors for each biquad section
          L1 = 0  (in the case of a single biquad)
          L1 = number of biquad sections
             (for multiple biquads)
          I4 -> scaled biquad coefficients
          L4 = 5 x [number of biquads]
          M0, M4 = 1
          M1 = -3
          M2 = 1 (in the case of multiple biquads)
          M2 = 0 (in the case of a single biquad)
          M3 = (1 - length of delay line buffer)
```

Software Examples 14

```
        Return Value:
            SR1 = output sample Y(n)

        Altered Registers:
            SE, MX0, MX1, MY0, MR, SR

        Computation Time (with N even):
            ADSP-2101/2102: (8 x N/2) + 5 cycles
            ADSP-2100/2100A: (8 x N/2) + 5 + 5 cycles

        All coefficients and data values are assumed to
        be in 1.15 format
}

.ENTRY      biquad;

biquad:     CNTR = number_of_biquads
            DO sections UNTIL CE;    {Loop once for each biquad}
                SE=DM(I1,M2);            {Scale factor for biquad}
                MX0=DM(I0,M0), MY0=PM(I4,M4);
                MR=MX0*MY0(SS), MX1=DM(I0,M0), MY0=PM(I4,M4);
                MR=MR+MX1*MY0(SS), MY0=PM(I4,M4);
                MR=MR+SR1*MY0(SS), MX0=DM(I0,M0), MY0=PM(I4,M4);
                MR=MR+MX0*MY0(SS), MX0=DM(I0,M1), MY0=PM(I4,M4);
                DM(I0,M0)=MX1, MR=MR+MX0*MY0(RND);
sections:       DM(I0,M0)=SR1, SR=ASHIFT MR1 (HI);
            DM(I0,M0)=MX0;
            DM(I0,M3)=SR1;
            RTS;
.ENDMOD;
```

Listing 14.2 Cascaded Biquad IIR Filter

14.5   SINE APPROXIMATION

The following formula approximates the sine of the input variable x:

$$\sin(x) = 3.140625x + 0.02026367x^2 - 5.325196x^3 + 0.5446778x^4 + 1.800293x^5$$

The approximation is accurate for any value of x from $0°$ to $90°$ (the first quadrant). However, because $\sin(-x) = -\sin(x)$ and $\sin(x) = \sin(180° - x)$, you can infer the sine of any angle from the sine of an angle in the first quadrant.

14 Software Examples

The routine that implements this sine approximation, accurate to within two LSBs, is shown in Listing 14.3. This routine accepts input values in 1.15 format. The coefficients, which are initialized in data memory in 4.12 format, have been adjusted to reflect an input value scaled to the maximum range allowed by this format. On this scale, 180° equals the maximum positive value, 0x7FFF, and –180° equals the maximum negative value, 0x8000.

The routine shown in Listing 14.3 first adjusts the input angle to its equivalent in the first quadrant. The sine of the modified angle is calculated by multiplying increasing powers of the angle by the appropriate coefficients. The result is adjusted if necessary to compensate for the modifications made to the original input value.

```
.MODULE   Sin_Approximation;

{
    Sine Approximation
          Y = Sin(x)

    Calling Parameters
          AX0 = x in scaled 1.15 format
          M3 = 1
          L3 = 0

    Return Values
          AR = y in 1.15 format

    Altered Registers
          AY0,AF,AR,MY1,MX1,MF,MR,SR,I3

    Computation Time
          25 cycles
}
```

Software Examples 14

```
.VAR/DM    sin_coeff[5];

.INIT      sin_coeff : 0x3240, 0x0053, 0xAACC, 0x08B7, 0x1CCE;

.ENTRY     sin;

sin:       I3=^sin_coeff;                  {Pointer to coeff. buffer}
           AY0=0x4000;
           AR=AX0, AF=AX0 AND AY0;         {Check 2nd or 4th quad.}
           IF NE AR=-AX0;                  {If yes, negate input}
           AY0=0x7FFF;
           AR=AR AND AY0;                  {Remove sign bit}
           MY1=AR;
           MF=AR*MY1 (RND), MX1=DM(I3,M3);  {MF = x²}
           MR=MX1*MY1 (SS), MX1=DM(I3,M3);  {MR = C₁x}
           CNTR=3;
           DO approx UNTIL CE;
              MR=MR+MX1*MF (SS);
approx:       MF=AR*MF (RND), MX1=DM(I3,M3);
           MR=MR+MX1*MF (SS);
           SR=ASHIFT MR1 BY 3 (HI);
           SR=SR OR LSHIFT MR0 BY 3 (LO);   {Convert to 1.15 format}
           AR=PASS SR1;
           IF LT AR=PASS AY0;              {Saturate if needed}
           AF=PASS AX0;
           IF LT AR=-AR;                   {Negate output if needed}
           RTS;
.ENDMOD;
```

Listing 14.3 Sine Approximation

14.6   SINGLE-PRECISION MATRIX MULTIPLY

The routine presented in this section multiplies two input matrices: X, an
RxS (R rows, S columns) matrix stored in data memory and Y, an SxT
(S rows, T columns) matrix stored in program memory. The output Z, an
RxT (R rows, T columns) matrix, is written to data memory.

The routine is shown in Listing 14.4. It requires a number of registers to be
initialized, as listed in the Calling Parameters section of the initial
comment. SE must contain the value necessary to shift the result of each
multiplication into the desired format. For example, SE would be set to
zero to obtain a matrix of 1.31 values from the multiplication of two
matrices of 1.15 values.

14 Software Examples

```
.MODULE     matmul;

{

            Single-Precision Matrix Multiplication

                        S
            Z(i,j) = Σ [X(i,k) × Y(k,j)]    i=0 to R;  j=0 to T
                     k=0

            X is an RxS matrix
            Y is an SxT matrix
            Z is an RxT matrix

        Calling Parameters
            I1 —> Z buffer in data memory                 L1 = 0
            I2 —> X, stored by rows in data memory        L2 = 0
            I6 —> Y, stored by rows in program memory     L6 = 0
            M0 = 1          M1 = S
            M4 = 1          M5 = T
            L0,L4,L5 = 0
            SE = Appropriate scale value
            CNTR = R

        Return Values
            Z Buffer filled by rows

        Altered Registers
            I0,I1,I2,I4,I5,MR,MX0,MY0,SR

        Computation Time
            ((S + 8) × T + 4) × R + 2 + 2 cycles

}
```

```
.ENTRY        spmm;

spmm:    DO row_loop UNTIL CE;
             I5=I6;                           {I5 = start of Y}
             CNTR=M5;
             DO column_loop UNTIL CE;
                 I0=I2;                       {Set I0 to current X row}
                 I4=I5;                       {Set I4 to current Y col}
                 CNTR=M1;
                 MR=0, MX0=DM(I0,M0), MY0=PM(I4,M5); {Get 1st data}
                 DO element_loop UNTIL CE;
element_loop:        MR=MR+MX0*MY0 (SS), MX0=DM(I0,M0),
MY0=PM(I4,M5);
                 SR=ASHIFT MR1 (HI), MY0=DM(I5,M4);   {Update I5}
                 SR=SR OR LSHIFT MR0 (LO);            {Finish shift}
column_loop:     DM(I1,M0)=SR1;                       {Save output}
row_loop:    MODIFY(I2,M1);                   {Update I2 to next X row}
         RTS;
.ENDMOD;
```

Listing 14.4 Single-Precision Matrix Multiply

14.7   RADIX-2 DECIMATION-IN-TIME FFT

The FFT program includes three subroutines. The first subroutine scrambles the input data (places the data in bit-reversed address order), so that the FFT output will be in the normal, sequential order. The next subroutine computes the FFT and the third scales the output data to maintain the block floating-point data format.

The program is contained in four modules. The main module declares and initializes data buffers and calls subroutines. The other three modules contain the FFT, bit reversal, and block floating-point scaling subroutines. The main module calls the FFT and bit reversal subroutines. The FFT module calls the data scaling subroutine.

The FFT is performed in place; that is, the outputs are written to the same buffer that the inputs are read from.

14.7.1   Main Module

The dit_fft_main module is shown in Listing 14.5. N is the number of points in the FFT (in this example, N=1024) and N_div_2 is used for specifying the lengths of buffers. To change the number of points in the FFT, you change the value of these constants and the twiddle factors.

14 Software Examples

The data buffers twid_real and twid_imag in program memory hold the twiddle factor cosine and sine values. The inplacereal, inplaceimag, inputreal and inputimag buffers in data memory store real and imaginary data values. Sequentially ordered input data is stored in inputreal and inputimag. This data is scrambled and written to inplacereal and inplaceimag. A four-location buffer called padding is placed at the end of inplaceimag to allow data accesses to exceed the buffer length. This buffer assists in debugging but is not necessary in a real system. Variables (one-location buffers) named groups, bflys_per_group, node_space and blk_exponent are declared last.

The real parts (cosine values) of the twiddle factors are stored in the buffer twid_real. This buffer is initialized from the file twid_real.dat. Likewise, twid_imag.dat values initialize the twid_imag buffer that stores the sine values of the twiddle factors. In an actual system, the hardware would be set up to initialize these memory locations.

The variable called groups is initialized to N_div_2, and bflys_per_group and node_space are each initialized to 2 because there are two butterflies per group in the second stage of the FFT. The blk_exponent variable is initialized to zero. This exponent value is updated when the output data is scaled.

After the initializations are complete, two subroutines are called. The first subroutine places the input sequence in bit-reversed order. The second performs the FFT and calls the block floating-point scaling routine.

```
.MODULE/ABS=4           dit_fft_main;
.CONST                  N=1024, N_div_2=512; {For 1024 points}
.VAR/PM/RAM/CIRC        twid_real [N_div_2];
.VAR/PM/RAM/CIRC        twid_imag [N_div_2];
.VAR/DM/RAM/ABS=0       inplacereal [N], inplaceimag [N], padding
[4];
.VAR/DM/RAM/ABS=H#1000  inputreal [N], inputimag [N];
.VAR/DM/RAM             groups, bflys_per_group, node_space,
                        blk_exponent;

.INIT      twid_real: <twid_real.dat>;
.INIT      twid_imag: <twid_imag.dat>;
.INIT      inputreal: <inputreal.dat>;
.INIT      inputimag: <inputimag.dat>;
.INIT      inplaceimag: <inputimag.dat>;
.INIT      groups: N_div_2;
```

Software Examples 14

```
.INIT        bflys_per_group: 2;
.INIT        node_space: 2;
.INIT        blk_exponent: 0;
.INIT        padding: 0,0,0,0;              {Zeros after inplaceimag}

.GLOBAL      twid_real, twid_imag;
.GLOBAL      inplacereal, inplaceimag;
.GLOBAL      inputreal, inputimag;
.GLOBAL      groups, bflys_per_group, node_space, blk_exponent;


.EXTERNAL    scramble, fft_strt;

             CALL scramble;                {subroutine calls}
             CALL fft_strt;
             TRAP;                         {halt program}
.ENDMOD;
```

Listing 14.5 Main Module, Radix-2 DIT FFT

14.7.2   DIT FFT Subroutine

The radix-2 DIT FFT routine is shown in Listing 14.6. The constants N and log2N are the number of points and the number of stages in the FFT, respectively. To change the number of points in the FFT, you modify these constants.

The first and last stages of the FFT are performed outside of the loop that executes all the other stages. Treating the first and last stages individually allows them to be executed faster. In the first stage, there is only one butterfly per group, so the butterfly loop is unnecessary, and the twiddle factors are all either 1 or 0, so no multiplications are necessary. In the last stage, there is only one group, so the group loop is unnecessary, as are the setup operations for the next stage.

14 Software Examples

```
{1024 point DIT radix 2 FFT}
{Block Floating Point Scaling}

.MODULE     fft;

{     Calling Parameters
            inplacereal=real input data in scrambled order
            inplaceimag=all zeroes (real input assumed)
            twid_real=twiddle factor cosine values
            twid_imag=twiddle factor sine values
            groups=N/2
            bflys_per_group=1
            node_space=1

      Return Values
            inplacereal=real FFT results, sequential order
            inplaceimag=imag. FFT results, sequential order

      Altered Registers
            I0,I1,I2,I3,I4,I5,L0,L1,L2,L3,L4,L5
            M0,M1,M2,M3,M4,M5
            AX0,AX1,AY0,AY1,AR,AF
            MX0,MX1,MY0,MY1,MR,SB,SE,SR,SI

      Altered Memory
            inplacereal, inplaceimag, groups, node_space,
            bflys_per_group, blk_exponent
}

.CONST      log2N=10, N=1024, nover2=512, nover4=256;

.EXTERNAL   twid_real, twid_imag;
.EXTERNAL   inplacereal, inplaceimag;
.EXTERNAL   groups, bflys_per_group, node_space;
.EXTERNAL   bfp_adj;
.ENTRY      fft_strt;

fft_strt:   CNTR=log2N - 2;    {Initialize stage counter}
            M0=0;
            M1=1;
            L1=0;
            L2=0;
            L3=0;
            L4=%twid_real;
            L5=%twid_imag;
            L6=0;
            SB=-2;
```

Software Examples 14

```
{———— STAGE 1 ————}

        I0=^inplacereal;
        I1=^inplacereal + 1;
        I2=^inplaceimag;
        I3=^inplaceimag + 1;
        M2=2;

        CNTR=nover2;
        AX0=DM(I0,M0);
        AY0=DM(I1,M0);
        AY1=DM(I3,M0);

        DO group_lp UNTIL CE;
            AR=AX0+AY0, AX1=DM(I2,M0);
            SB=EXPADJ AR, DM(I0,M2)=AR;
            AR=AX0-AY0;
            SB=EXPADJ AR;
            DM(I1,M2)=AR, AR=AX1+AY1;
            SB=EXPADJ AR, DM(I2,M2)=AR;
            AR=AX1-AY1, AX0=DM(I0,M0);
            SB=EXPADJ AR, DM(I3,M2)=AR;
            AY0=DM(I1,M0);
group_lp:   AY1=DM(I3,M0);
        CALL bfp_adj;

{——————STAGES 2 TO N-1——————————}

        DO stage_loop UNTIL CE;      {Compute all stages in FFT}
        I0=^inplacereal;             {I0 ->x0 in 1st grp of stage}
        I2=^inplaceimag;             {I2 ->y0 in 1st grp of stage}
        SI=DM(groups);
        SR=ASHIFT SI BY -1(LO);      {groups / 2}
        DM(groups)=SR0;              {groups=groups / 2}
        CNTR=SR0;                    {CNTR=group counter}
        M4=SR0;                      {M4=twiddle factor modifier}
        M2=DM(node_space);           {M2=node space modifier}
        I1=I0;
        MODIFY(I1,M2);               {I1 ->y0 of 1st grp in stage}
        I3=I2;
        MODIFY(I3,M2);               {I3 ->y1 of 1st grp in stage}
```

14 Software Examples

```
                DO group_loop UNTIL CE;
                    I4=^twid_real;                          {I4 -> C of W0}
                    I5=^twid_imag;                          {I5 -> (-S) of W0}
                    CNTR=DM(bflys_per_group);               {CNTR=bfly count}
                    MY0=PM(I4,M4),MX0=DM(I1,M0);            {MY0=C,MX0=x1 }
                    MY1=PM(I5,M4),MX1=DM(I3,M0);            {MY1=-S,MX1=y1}
                    DO bfly_loop UNTIL CE;
                        MR=MX0*MY1(SS),AX0=DM(I0,M0);
                                                            {MR=x1(-S),AX0=x0}
                        MR=MR+MX1*MY0(RND),AX1=DM(I2,M0);
                                                    {MR=(y1(C)+x1(-S)),AX1=y0}
                        AY1=MR1,MR=MX0*MY0(SS);
                                                    {AY1=y1(C)+x1(-S),MR=x1(C)}
                        MR=MR-MX1*MY1(RND);             {MR=x1(C)-y1(-S)}
                        AY0=MR1,AR=AX1-AY1;
                                {AY0=x1(C)-y1(-S),AR=y0-[y1(C)+x1(-S)]}
                        SB=EXPADJ AR,DM(I3,M1)=AR;
                            {Check for bit growth, y1=y0-[y1(C)+x1(-S)]}
                        AR=AX0-AY0,MX1=DM(I3,M0),MY1=PM(I5,M4);
                    {AR=x0-[x1(C)-y1(-S)], MX1=next y1,MY1=next (-S)}
                        SB=EXPADJ AR,DM(I1,M1)=AR;
                            {Check for bit growth, x1=x0-[x1(C)-y1(-S)]}
                        AR=AX0+AY0,MX0=DM(I1,M0),MY0=PM(I4,M4);
                          {AR=x0+[x1(C)-y1(-S)], MX0=next x1,MY0=next C}
                        SB=EXPADJ AR,DM(I0,M1)=AR;
                            {Check for bit growth, x0=x0+[x1(C)-y1(-S)]}
                        AR=AX1+AY1;                     {AR=y0+[y1(C)+x1(-S)]}
bfly_loop:              SB=EXPADJ AR,DM(I2,M1)=AR;
                            {Check for bit growth, y0=y0+[y1(C)+x1(-S)]}
                    MODIFY(I0,M2);              {I0 ->1st x0 in next group}
                    MODIFY(I1,M2);              {I1 ->1st x1 in next group}
                    MODIFY(I2,M2);              {I2 ->1st y0 in next group}
group_loop:         MODIFY(I3,M2);              {I3 ->1st y1 in next group}

                CALL bfp_adj;                      {Compensate for bit growth}
                SI=DM(bflys_per_group);
                SR=ASHIFT SI BY 1(LO);
                DM(node_space)=SR0;                {node_space=node_space / 2}
stage_loop: DM(bflys_per_group)=SR0;
                            {bflys_per_group=bflys_per_group / 2}
```

Software Examples 14

```
{———— LAST STAGE ————}

        I0=^inplacereal;
        I1=^inplacereal+nover2;
        I2=^inplaceimag;
        I3=^inplaceimag+nover2;

        CNTR=nover2;
        M2=DM(node_space);
        M4=1;
        I4=^twid_real;
        I5=^twid_imag;

        MY0=PM(I4,M4),MX0=DM(I1,M0);              {MY0=C,MX0=x1}
        MY1=PM(I5,M4),MX1=DM(I3,M0);              {MY1=-S,MX1=y1}
        DO bfly_lp UNTIL CE;
            MR=MX0*MY1(SS),AX0=DM(I0,M0);           {MR=x1(-S),AX0=x0}
            MR=MR+MX1*MY0(RND),AX1=DM(I2,M0);
                                          {MR=(y1(C)+x1(-S)),AX1=y0}
            AY1=MR1,MR=MX0*MY0(SS);      {AY1=y1(C)+x1(-S),MR=x1(C)}
            MR=MR-MX1*MY1(RND);                       {MR=x1(C)-y1(-S)}
            AY0=MR1,AR=AX1-AY1;
                        {AY0=x1(C)-y1(-S), AR=y0-[y1(C)+x1(-S)]}
            SB=EXPADJ AR,DM(I3,M1)=AR;
                    {Check for bit growth, y1=y0-[y1(C)+x1(-S)]}
            AR=AX0-AY0,MX1=DM(I3,M0),MY1=PM(I5,M4);
                  {AR=x0-[x1(C)-y1(-S)], MX1=next y1,MY1=next (-S)}
            SB=EXPADJ AR,DM(I1,M1)=AR;
                    {Check for bit growth, x1=x0-[x1(C)-y1(-S)]}
            AR=AX0+AY0,MX0=DM(I1,M0),MY0=PM(I4,M4);
                    {AR=x0+[x1(C)-y1(-S)], MX0=next x1,MY0=next C}
            SB=EXPADJ AR,DM(I0,M1)=AR;
                    {Check for bit growth, x0=x0+[x1(C)-y1(-S)]}
            AR=AX1+AY1;                       {AR=y0+[y1(C)+x1(-S)]}
bfly_lp:    SB=EXPADJ AR,DM(I2,M1)=AR;        {Check for bit growth}

        CALL bfp_adj;

        RTS;
.ENDMOD;
```

Listing 14.6 Radix-2 DIT FFT Routine, Conditional Block Floating-Point

14 Software Examples

14.7.3 Bit-Reverse Subroutine

The bit-reversal routine, called scramble, puts the input data in bit-reversed order so that the results will be in sequential order. This routine uses the bit-reverse capability of the ADSP-2100 family processors.

```
.MODULE  dit_scramble;

{  Calling Parameters
        Sequentially ordered input data in inputreal

    Return Values
        Scrambled input data in inplacereal

    Altered Registers
        I0,I4,M0,M4,AY1

    Altered Memory
        inplacereal
}

.CONST      N=1024,mod_value=H#0010; {Initialize constants}

.EXTERNAL   inputreal, inplacereal;

.ENTRY      scramble;


scramble:   I4=^inputreal;   {I4->sequentially ordered data}
            I0=^inplacereal;     {I0->scrambled data}
            M4=1;
            M0=mod_value;    {M0=modifier for reversing N bits}
            L4=0;
            L0=0;
            CNTR = N;
            ENA BIT_REV;     {Enable bit-reversed outputs on DAG1}
            DO brev UNTIL CE;
               AY1=DM(I4,M4);    {Read sequentially ordered data}
brev:          DM(I0,M0)=AY1;
                            {Write data in bit-reversed location}
            DIS BIT_REV;     {Disable bit-reverse}
            RTS;             {Return to calling program}
.ENDMOD;
```

Listing 14.7 Bit-Reverse Routine (Scramble)

Software Examples 14

14.7.4 Block Floating-Point Scaling Subroutine

The bfp_adj routine checks the FFT output data for bit growth and scales the entire set of data if necessary. This check prevents data overflow for each stage in the FFT. The routine, shown in Listing 14.8, uses the exponent detection capability of the shifter.

```
.MODULE   dit_radix_2_bfp_adjust;

{  Calling Parameters
       Radix-2 DIT FFT stage results in inplacereal and inplaceimag

   Return Parameters
       inplacereal and inplaceimag adjusted for bit growth

   Altered Registers
       I0,I1,AX0,AY0,AR,MX0,MY0,MR,CNTR

   Altered Memory
       inplacereal, inplaceimag, blk_exponent
}

.CONST      Ntimes2 = 2048;
.EXTERNAL   inplacereal, blk_exponent;  {Begin declaration section}

.ENTRY      bfp_adj;

bfp_adj:    AY0=CNTR;               {Check for last stage}
            AR=AY0-1
            IF EQ RTS;              {If last stage, return}
            AY0=-2;
            AX0=SB;
            AR=AX0-AY0;             {Check for SB=-2}
            IF EQ RTS;              {IF SB=-2, no bit growth, return}
            I0=^inplacereal;        {I0=read pointer}
            I1=^inplacereal;        {I1=write pointer}
            AY0=-1;
            MY0=H#4000;             {Set MY0 to shift 1 bit right}
            AR=AX0-AY0,MX0=DM(I0,M1);
                                    {Check if SB=-1; Get 1st sample}
```

371

14 Software Examples

```
                IF EQ JUMP strt_shift;
                                {If SB=-1, shift block data 1 bit}
                AX0=-2;             {Set AX0 for block exponent update}
                MY0=H#2000;         {Set MY0 to shift 2 bits right}
strt_shift:  CNTR=Ntimes2 - 1;        {initialize loop counter}
                DO shift_loop UNTIL CE;     {Shift block of data}
                  MR=MX0*MY0(RND),MX0=DM(I0,M1);
                                {MR=shifted data,MX0=next value}
shift_loop:      DM(I1,M1)=MR1;     {Unshifted data=shifted data}
                MR=MX0*MY0(RND);          {Shift last data word}
                AY0=DM(blk_exponent);      {Update block exponent and}
                DM(I1,M1)=MR1,AR=AY0-AX0;  {store last shifted sample}
                DM(blk_exponent)=AR;
                RTS;
.ENDMOD;
```

Listing 14.8 Radix-2 Block Floating-Point Scaling Routine

Instruction Coding ■ A

A.1 OPCODES

This appendix gives a summary of the complete instruction set of the ADSP-2100 family processors. Opcode field names are defined at the end of the appendix. Any instruction codes not shown are reserved for future use.

Type 1: ALU / MAC with Data & Program Memory Read

23	22	21 20	19 18	17 16 15 14 13 12	11 10 9	8 7 6	5	4	3	2 1 0
1	1	PD	DD	AMF	Yop	Xop	PM I	PM M	DM I	DM M

Type 2: Data Memory Write (Immediate Data)

23	22	21	20	19 18 17 16 15 14 13 12 11 10 9 8 7 6 5 4 3	2 1	0
1	0	1	G	DATA	I	M

Type 3: Read /Write Data Memory (Immediate Address)

23	22	21	20	19 18	17 16 15 14 13 12 11 10 9 8 7 6 5	4 3 2 1 0
1	0	0	D	RGP	ADDR	REG

Type 4: ALU / MAC with Data Memory Read / Write

23	22	21	20	19	18	17 16 15 14 13 12	11 10 9	8 7 6	5 4 3	2 1	0
0	1	1	G	D	Z	AMF	Yop	Xop	DREG	I	M

Type 5: ALU / MAC with Program Memory Read / Write

23	22	21	20	19	18	17 16 15 14 13 12	11 10 9	8 7 6	5 4 3	2 1	0
0	1	0	1	D	Z	AMF	Yop	Xop	DREG	I	M

A Instruction Coding

Type 6: Load Data Register Immediate

23 22 21 20	19 18 17 16 15 14 13 12 11 10 9 8 7 6 5 4 3	2 1 0
0 1 0 0	DATA	DREG

Type 7: Load Non-Data Register Immediate

23 22 21 20	19 18	17 16 15 14 13 12 11 10 9 8 7 6 5 4 3	2 1 0
0 0 1 1	RGP	DATA	REG

Type 8: ALU / MAC with Internal Data Register Move

23 22 21 20 19	18	17 16 15 14 13	12 11 10	9 8	7 6 5 4	3 2 1 0
0 0 1 0 1	Z	AMF	Yop	Xop	Dest DREG	Source DREG

Type 9: Conditional ALU / MAC

23 22 21 20 19	18	17 16 15 14 13	12 11 10	9 8	7 6 5 4	3 2 1 0
0 0 1 0 0	Z	AMF	Yop	Xop	0 0 0 0	COND

Type 10: Conditional Jump (Immediate Address)

23 22 21 20 19	18	17 16 15 14 13 12 11 10 9 8 7 6 5 4 3	2 1 0
0 0 0 1 1	S	ADDR	COND

Type 11: Do Until

23 22 21 20 19 18	17 16 15 14 13 12 11 10 9 8 7 6 5 4 3	2 1 0
0 0 0 1 0 1	ADDR	TERM

Type 12: Shift with Data Memory Read / Write

23 22 21 20 19 18 17	16	15	14 13 12 11	10 9 8	7 6 5 4 3	2 1	0
0 0 0 1 0 0 1	G	D	SF	Xop	DREG	I	M

Instruction Coding   A

Type 13:   Shift with Program Memory Read / Write

23	22	21	20	19	18	17	16	15	14	13	12	11	10	9	8	7	6	5	4	3	2	1	0
0	0	0	1	0	0	0	1	D		SF			Xop			DREG				I		M	

Type 14:   Shift with Internal Data Register Move

23	22	21	20	19	18	17	16	15	14	13	12	11	10	9	8	7	6	5	4	3	2	1	0
0	0	0	1	0	0	0	0	0			SF			Xop		Dest DREG				Source DREG			

Type 15:   Shift Immediate

23	22	21	20	19	18	17	16	15	14	13	12	11	10	9	8	7	6	5	4	3	2	1	0
0	0	0	0	1	1	1	1	0			SF			Xop				exponent					

Type 16:   Conditional Shift

23	22	21	20	19	18	17	16	15	14	13	12	11	10	9	8	7	6	5	4	3	2	1	0
0	0	0	0	1	1	1	0	0			SF			Xop		0	0	0	0	COND			

Type 17:   Internal Data Move

23	22	21	20	19	18	17	16	15	14	13	12	11	10	9	8	7	6	5	4	3	2	1	0
0	0	0	0	1	1	0	1	0	0	0	0	DST RGP		SRC RGP		Dest REG			Source REG				

Type 18:   Mode Control

23	22	21	20	19	18	17	16	15	14	13	12	11	10	9	8	7	6	5	4	3	2	1	0
0	0	0	0	1	1	0	0	TI		MM		AS		OL		BR		SR		GM		0	0

Definitions for the field names shown (TI, MM, AS, OL, BR, SR, GM) can be found under "Mode Control Codes" at the back of this appendix.

A Instruction Coding

Type 19: Conditional Jump (Indirect Address)

23	22	21	20	19	18	17	16	15	14	13	12	11	10	9	8	7	6	5	4	3	2	1	0
0	0	0	0	1	0	1	1	0	0	0	0	0	0	0	0	I	0	S	COND				

Type 20: Conditional Return

23	22	21	20	19	18	17	16	15	14	13	12	11	10	9	8	7	6	5	4	3	2	1	0
0	0	0	0	1	0	1	0	0	0	0	0	0	0	0	0	0	0	0	T	COND			

Type 21: Modify Address Register

23	22	21	20	19	18	17	16	15	14	13	12	11	10	9	8	7	6	5	4	3	2	1	0
0	0	0	0	1	0	0	1	0	0	0	0	0	0	0	0	0	0	G	I	M			

Type 22: Conditional TRAP (ADSP-2100 Only)

23	22	21	20	19	18	17	16	15	14	13	12	11	10	9	8	7	6	5	4	3	2	1	0
0	0	0	0	1	0	0	0	0	0	0	0	0	0	0	0	0	0	0	0	COND			

Type 23: DIVQ

23	22	21	20	19	18	17	16	15	14	13	12	11	10	9	8	7	6	5	4	3	2	1	0
0	0	0	0	0	1	1	1	0	0	0	1	0	Xop			0	0	0	0	0	0	0	0

Type 24: DIVS

23	22	21	20	19	18	17	16	15	14	13	12	11	10	9	8	7	6	5	4	3	2	1	0
0	0	0	0	0	1	1	0	0	0	0	Yop			Xop			0	0	0	0	0	0	0

Type 25: Saturate MR

23	22	21	20	19	18	17	16	15	14	13	12	11	10	9	8	7	6	5	4	3	2	1	0
0	0	0	0	0	1	0	1	0	0	0	0	0	0	0	0	0	0	0	0	0	0	0	0

Instruction Coding   A

Type 26:   Stack Control

23	22	21	20	19	18	17	16	15	14	13	12	11	10	9	8	7	6	5	4	3	2	1	0
0	0	0	0	0	1	0	0	0	0	0	0	0	0	0	0	0	0	0	0	PP	LP	CP	SPP

Type 27:   Call or Jump on Flag In (Not ADSP-2100)

23	22	21	20	19	18	17	16	15	14	13	12	11	10	9	8	7	6	5	4	3	2	1	0
0	0	0	0	0	0	1	1				Address									Addr	FIC		S

                                    12 LSBs   2 MSBs

Type 28:   Flag Out Mode Control (Not ADSP-2100)

23	22	21	20	19	18	17	16	15	14	13	12	11	10	9	8	7	6	5	4	3	2	1	0
0	0	0	0	0	0	1	0	0	0	0	0	0	0	0	0	0	0	FO		COND			

Type 29:   Reserved

23	22	21	20	19	18	17	16	15	14	13	12	11	10	9	8	7	6	5	4	3	2	1	0
0	0	0	0	0	0	0	1	x	x	x	x	x	x	x	x	x	x	x	x	x	x	x	x

Type 30:   No Operation (NOP)

23	22	21	20	19	18	17	16	15	14	13	12	11	10	9	8	7	6	5	4	3	2	1	0
0	0	0	0	0	0	0	0	0	0	0	0	0	0	0	0	0	0	0	0	0	0	0	0

Type 31:   Idle (Not ADSP-2100)

23	22	21	20	19	18	17	16	15	14	13	12	11	10	9	8	7	6	5	4	3	2	1	0
0	0	0	0	0	0	1	0	1	0	0	0	0	0	0	0	0	0	0	0	0	0	0	0

A Instruction Coding

A.2 ABBREVIATION CODING

AMF   ALU / MAC Function codes

0 0 0 0 0   No operation

MAC Function codes

0	0	0	0	1	X * Y	(RND)		
0	0	0	1	0	MR + X * Y	(RND)		
0	0	0	1	1	MR − X * Y	(RND)		
0	0	1	0	0	X * Y	(SS)	Clear when y = 0
0	0	1	0	1	X * Y	(SU)	
0	0	1	1	0	X * Y	(US)	
0	0	1	1	1	X * Y	(UU)	
0	1	0	0	0	MR + X * Y	(SS)	
0	1	0	0	1	MR + X * Y	(SU)	
0	1	0	1	0	MR + X * Y	(US)	
0	1	0	1	1	MR + X * Y	(UU)	
0	1	1	0	0	MR − X * Y	(SS)	
0	1	1	0	1	MR − X * Y	(SU)	
0	1	1	1	0	MR − X * Y	(US)	
0	1	1	1	1	MR − X * Y	(UU)	

ALU Function codes

1	0	0	0	0	Y	Clear when y = 0
1	0	0	0	1	Y + 1	PASS 1 when y = 0
1	0	0	1	0	X + Y + C	
1	0	0	1	1	X + Y	X when y = 0
1	0	1	0	0	NOT Y	
1	0	1	0	1	− Y	
1	0	1	1	0	X − Y + C − 1	X + C − 1 when y = 0
1	0	1	1	1	X − Y	
1	1	0	0	0	Y − 1	PASS −1 when y = 0
1	1	0	0	1	Y − X	− X when y = 0
1	1	0	1	0	Y − X + C − 1	−X + C − 1 when y = 0
1	1	0	1	1	NOT X	

Instruction Coding   A

1	1	1	0	0	X AND Y
1	1	1	0	1	X OR Y
1	1	1	1	0	X XOR Y
1	1	1	1	1	ABS X

COND   Status Condition codes

0	0	0	0	Equal	EQ
0	0	0	1	Not equal	NE
0	0	1	0	Greater than	GT
0	0	1	1	Less than or equal	LE
0	1	0	0	Less than	LT
0	1	0	1	Greater than or equal	GE
0	1	1	0	ALU Overflow	AV
0	1	1	1	NOT ALU Overflow	NOT AV
1	0	0	0	ALU Carry	AC
1	0	0	1	Not ALU Carry	NOT AC
1	0	1	0	X input sign negative	NEG
1	0	1	1	X input sign positive	POS
1	1	0	0	MAC Overflow	MV
1	1	0	1	Not MAC Overflow	NOT MV
1	1	1	0	Not counter expired	NOT CE
1	1	1	1	Always true	

CP   Counter Stack Pop codes

0	No change
1	Pop

D   Memory Access Direction codes

0	Read
1	Write

A Instruction Coding

DD   Double Data Fetch Data Memory Destination codes

0	0	AX0
0	1	AX1
1	0	MX0
1	1	MX1

DREG   Data Register codes

0	0	0	0	AX0
0	0	0	1	AX1
0	0	1	0	MX0
0	0	1	1	MX1
0	1	0	0	AY0
0	1	0	1	AY1
0	1	1	0	MY0
0	1	1	1	MY1
1	0	0	0	SI
1	0	0	1	SE
1	0	1	0	AR
1	0	1	1	MR0
1	1	0	0	MR1
1	1	0	1	MR2
1	1	1	0	SR0
1	1	1	1	SR1

FIC   FI condition code

1	latched FI is 1	FLAG_IN
0	latched FI is 0	NOT FLAG_IN

Instruction Coding   A

FO   Mode Control codes for Flag Out pin

       FO: Set, reset, or toggle the output Flag.

0 0	No change
0 1	Toggle
1 0	Reset
1 1	Set

G   Data Address Generator codes

0	DAG1
1	DAG2

I   Index Register codes

G =	0	1
0 0	I0	I4
0 1	I1	I5
1 0	I2	I6
1 1	I3	I7

LP   Loop Stack Pop codes

0	No Change
1	Pop

M   Modify Register codes

G =	0	1
0 0	M0	M4
0 1	M1	M5
1 0	M2	M6
1 1	M3	M7

A Instruction Coding

Mode Control codes

SR:	Secondary register bank
BR:	Bit-reverse mode
OL:	ALU overflow latch mode
AS:	AR register saturate mode
MM:	Alternate Multiplier placement mode (not ADSP-2100)
GM:	GOMode; enable means go if possible (not ADSP-2100)
TI:	Timer enable (not ADSP-2100)

0 0	No change
0 1	No change
1 0	Disable
1 1	Enable

PD   Double Data Fetch Program Memory Destination codes

0 0	AY0
0 1	AY1
1 0	MY0
1 1	MY1

PP   PC Stack Pop codes

0	No Change
1	Pop

Instruction Coding   A

REG   Register codes
Codes not assigned are reserved for future use.

RGP =		00	01	10	11
0 0 0 0		AX0	I0	I4	ASTAT
0 0 0 1		AX1	I1	I5	MSTAT
0 0 1 0		MX0	I2	I6	SSTAT (read only)
0 0 1 1		MX1	I3	I7	IMASK
0 1 0 0		AY0	M0	M4	ICNTL
0 1 0 1		AY1	M1	M5	CNTR
0 1 1 0		MY0	M2	M6	SB
0 1 1 1		MY1	M3	M7	PX
1 0 0 0		SI	L0	L4	RX0
1 0 0 1		SE	L1	L5	TX0
1 0 1 0		AR	L2	L6	RX1 *not ADSP-2100 registers*
1 0 1 1		MR0	L3	L7	TX1
1 1 0 0		MR1	–	–	IFC (write only)
1 1 0 1		MR2	–	–	OWRCNTR (write only)
1 1 1 0		SR0	–	–	–
1 1 1 1		SR1	–	–	–

S   Jump/Call codes

0	Jump
1	Call

A   Instruction Coding

SF   Shifter Function codes

0	0	0	0	LSHIFT	(HI)
0	0	0	1	LSHIFT	(HI, OR)
0	0	1	0	LSHIFT	(LO)
0	0	1	1	LSHIFT	(LO, OR)
0	1	0	0	ASHIFT	(HI)
0	1	0	1	ASHIFT	(HI, OR)
0	1	1	0	ASHIFT	(LO)
0	1	1	1	ASHIFT	(LO, OR)
1	0	0	0	NORM	(HI)
1	0	0	1	NORM	(HI, OR)
1	0	1	0	NORM	(LO)
1	0	1	1	NORM	(LO, OR)
1	1	0	0	EXP	(HI)
1	1	0	1	EXP	(HIX)
1	1	1	0	EXP	(LO)
1	1	1	1	Derive Block Exponent	

SPP   Status Stack Push/Pop codes

0	0	No change
0	1	No change
1	0	Push
1	1	Pop

T   Return Type codes

0	Return from Subroutine
1	Return from Interrupt

Instruction Coding   A

TERM   Termination codes for DO UNTIL

0	0	0	0	Not equal	NE
0	0	0	1	Equal	EQ
0	0	1	0	Less than or equal	LE
0	0	1	1	Greater than	GT
0	1	0	0	Greater than or equal	GE
0	1	0	1	Less than	LT
0	1	1	0	NOT ALU Overflow	NOT AV
0	1	1	1	ALU Overflow	AV
1	0	0	0	Not ALU Carry	NOT AC
1	0	0	1	ALU Carry	AC
1	0	1	0	X input sign positive	POS
1	0	1	1	X input sign negative	NEG
1	1	0	0	Not MAC Overflow	NOT MV
1	1	0	1	MAC Overflow	MV
1	1	1	0	Counter expired	CE
1	1	1	1	Always	FOREVER

X   X Operand codes

0	0	0	X0 (SI for shifter)
0	0	1	X1 (invalid for shifter)
0	1	0	AR
0	1	1	MR0
1	0	0	MR1
1	0	1	MR2
1	1	0	SR0
1	1	1	SR1

A   Instruction Coding

Y   Y Operand codes

```
0 0          Y0
0 1          Y1
1 0          F (feedback register)
1 1          zero
```

Z   ALU/MAC Result Register codes

```
0            Result register
1            Feedback register
```

Division Exceptions ■ B

B.1 DIVISION FUNDAMENTALS

The ADSP-2100 family processors' instruction set contains two instructions for implementing a non-restoring divide algorithm. These instructions take as their operands twos-complement or unsigned numbers, and in sixteen cycles produce a truncated quotient of sixteen bits. For most numbers and applications, these primitives produce the correct results. However, there are certain situations where results produced will be off by one LSB. This appendix documents these situations, and presents alternatives for producing the correct results.

Computing a 16-bit fixed point quotient from two numbers is accomplished by 16 executions of the DIVQ instruction for unsigned numbers. Signed division uses the DIVS instruction first, followed by fifteen DIVQs. Regardless of which division you perform, both input operands must be of the same type (signed or unsigned) and produce a result of the same type.

These two instructions are used to implement a conditional add/subtract, non-restoring division algorithm. As its name implies, the algorithm functions by adding or subtracting the divisor to/from the dividend. The decision as to which operation is perform is based on the previously generated quotient bit. Each add/subtract operation produces a new partial remainder, which will be used in the next step.

The phrase non-restoring refers to the fact that the final remainder is not correct. With a restoring algorithm, it is possible, at any step, to take the partial quotient, multiply it by the divisor, and add the partial remainder to recreate the dividend. With this non-restoring algorithm, it is necessary to add two times the divisor to the partial remainder if the previously determined quotient bit is zero. It is easier to compute the remainder using the multiplier than in the ALU.

B.1.1 Signed Division

Signed division is accomplished by first storing the 16-bit divisor in an X register (AX0, AX1, AR, MR2, MR1, MR0, SR1, or SR0). The 32-bit dividend must be stored in two separate 16-bit registers. The lower 16-bits must be stored in AY0, while the upper 16-bits can be in either AY1, or AF.

387

B   Division Exceptions

The DIVS primitive is executed once, with the proper operands (ex. DIVS AY1, AX0) to compute the sign of the quotient. The sign bit of the quotient is determined by XORing (exclusive-or) the sign bits of each operand. The entire 32-bit dividend is shifted left one bit. The lower fifteen bits of the dividend with the recently determined sign bit appended are stored in AY0, while the lower fifteen bits of the upper word, with the MSB of the lower word appended is stored in AF.

To complete the division, 15 DIVQ instructions are executed. Operation of the DIVQ primitive is described below.

B.1.2   Unsigned Division

Computing an unsigned division is done like signed division, except the first instruction is not a DIVS, but another DIVQ. The upper word of the dividend must be stored in AF, and the AQ bit of the ASTAT register must be set to zero before the divide begins.

The DIVQ instruction uses the AQ bit of the ASTAT register to determine if the dividend should be added to, or subtracted from the partial remainder stored in AF and AY0. If AQ is zero, a subtract occurs. A new value for AQ is determined by XORing the MSB of the divisor with the MSB of the dividend. The 32-bit dividend is shifted left one bit, and the inverted value of AQ is moved into the LSB.

B.1.3   Output Formats

As in multiplication, the format of a division result is based on the format of the input operands. The division logic has been designed to work most efficiently with fully fractional numbers, those most commonly used in fixed-point DSP applications. A signed, fully fractional number uses one bit before the binary point as the sign, with fifteen (or thirty-one in double precision) bits to the right, for magnitude.

If the dividend is in M.N format (M bits before the binary point, N bits after), and the divisor is O.P format, the quotient's format will be (M-O+1).(N-P-1). As you can see, dividing a 1.31 number by a 1.15 number will produce a quotient whose format is (1-1+1).(31-15-1) or 1.15.

Before dividing two numbers, you must ensure that the format of the quotient will be valid. For example, if you attempted to divide a 32.0 number by a 1.15 number the result would attempt to be in (32-1+1).(0-15-1) or 32.-16 format. This cannot be represented in a 16-bit register!

Division Exceptions   B

In addition to proper output format, you must insure that a divide overflow does not occur. Even if a division of two numbers produces a legal output format, it is possible that the number will overflow, and be unable to fit within the constraints of the output. For example, if you wished to divide a 16.16 number by a 1.15 number, the output format would be (16-1+1).(16-15-1) or 16.0 which is legal. Now assume you happened to have 16384 (H#4000) as the dividend and .25 (H#2000) as the divisor, the quotient would be 65536, which does not fit in 16.0 format. This operation would overflow, producing an erroneous results.

Input operands can be checked before division to ensure that an overflow will not result. If the magnitude of the upper 16 bits of the dividend is larger than the magnitude of the divisor, an overflow will result.

B.1.4   Integer Division

One special case of division that deserves special mention is integer division. There may be some cases where you wish to divide two integers, and produce an integer result. It can be seen that an integer-integer division will produce an invalid output format of (32-16+1).(0-0-1), or 17.-1.

To generate an integer quotient, you must shift the dividend to the left one bit, placing it in 31.1 format. The output format for this division will be (31-16+1).(1-0-1), or 16.0. You must ensure that no significant bits are lost during the left shift, or an invalid result will be generated.

B.2   ERROR SITUATIONS

Although the divide primitives for the ADSP-2100 family work in most instances, there are two cases where an invalid, or inaccurate result can be generated. The first case involves signed division by a negative number. If you attempt to use a negative number as the divisor, the quotient generated may be one LSB less than the correct result. The other case concerns unsigned division by a divisor greater than h#7FFF. If the divisor in an unsigned division exceeds H#7FFF, an invalid quotient will be generated.

B.2.1   Negative Divisor Error

The quotient produced during a divide involving a negative divisor will generally be one LSB less than the correct result. The divide algorithm implemented on the ADSP-2100 family does not correctly compensate for the twos-complement format of a negative number, causing this inaccuracy.

B   Division Exceptions

There is one case where this discrepancy does not occur. If the result of the division operation should equal H#8000, then it will be correctly represented, and not be one LSB off.

There are several ways to correct for this error. But before changing any code, you should determine if one LSB error in you quotient is significant problem. In some cases, the LSB is small enough to be insignificant. If you find it necessary have exact results, two solutions are apparent.

One way would be to avoid division by a negative number. If your divisor is negative, take its absolute value, and invert the sign of the quotient after division. This will produce the correct result.

Another technique would be to check the result by multiplying the quotient by the divisor. Compare this value with the dividend, if they are off by more than the value of the divisor, increase the quotient by one.

B.2.2   Unsigned Division Error

Unsigned divisions can produce erroneous results if the divisor is greater than H#7FFF. You should not attempt to divide two unsigned numbers if the divisor has a one in the MSB. If it is necessary to perform a such a division, both operands should be shifted right one bit. This will maintain the correct orientation of operands.

Shifting both operands may result in a one LSB error in the quotient. This can be solved by multiplying the quotient by the original (not shifted) divisor. Subtract this value from the original dividend to calculate the error. If the error is greater than the divisor, add one to the quotient, if it is negative, subtract one from the quotient.

B.3   SOFTWARE SOLUTION

Each of the problems mentioned in this Appendix can be compensated for in software. Listing 1 shows the module *divide_solution*. This code can be used to divide two signed or unsigned numbers to produce the correct quotient, or an error condition.

In addition to correcting the problems mentioned, this module provides a check for division overflow and computes the remainder following the division.

Division Exceptions   B

Since many applications do not require complete error checking, the code has been designed so you can remove tests that are not necessary for your project. This will decrease memory requirements, as well as increase execution speed.

The module *signed_div* expects the 32-bit dividend to be stored in AY1&AY0, and the divisor in AX0. Upon return either the AR register holds the quotient and MR0 holds the remainder, or the overflow flag is set. The entire routine takes at most twenty-seven cycles to execute. If an exception condition exists, it may return sooner. The first two instructions store the dividend in the MR registers, the absolute value of the dividend's MSW in AF, and the divisor's absolute value in AR.

The code block labeled *test_1* checks for division by H#8000. Attempting to take the absolute value of H#8000 produces an overflow. If the AV flag is set (from taking the absolute value of the divisor), then the quotient is −AY1. This can produce an error if AY1 is H#8000, so after taking the negative of AY1, the overflow flag is checked again. If it is set control is returned to the calling routine, otherwise the remainder is computed. If it is not necessary to check for a divisor of H#8000, this code block can be removed.

The code block labeled *test_2* checks for a division overflow condition. The absolute value of the divisor is subtracted from the absolute value of the dividend's MSW. If the divisor is less then the dividend, it is likely an overflow will occur. If the two are equal in magnitude, but different in sign, the result will be H#8000, so this special case is checked. If your application does not require an overflow check, this code block can be removed. If you decide to remove *test_2* be sure to change the JUMP address in *test_1* to *do_divs*, instead of *test_2*.

After error checking, the actual division is performed. Since the absolute value of the divisor has been stored in AR, this is used as the X-operand for the DIVS instruction. 15 DIVQ instructions follow, computing the rest of the quotient. The correct sign for the quotient is determined, based on the AS flag of the ASTAT register. Since the MR register contains the original dividend, the remainder can be determine by a multiply subtract operation. The divisor times the quotient is subtracted from MR to produce the remainder in MR0.

The last step before returning is to clear the ASTAT register which may contain an overflow flag produced during the divide.

B Division Exceptions

The subroutine *unsigned_div* is very similar to *signed_div*. MR1 and AF are loaded with the MSW of the dividend, MR0 is loaded with the dividend LSW and the divisor is passed into AR. Since unsigned division with a large divisor (>H#7FFF) is prohibited, the MSB of the divisor is checked. If it contains a one, the overflow flag is set, and the routine returns to the caller. Otherwise *test_11* checks for a standard divide overflow.

In *test_11* the divisor is subtracted from the MSW of the dividend. If the result is less then zero division can proceed, otherwise the overflow flag is set. If you wish to remove *test_11*, be sure to change the JUMP address in *test_10* to *do_divq*.

The actual unsigned division is performed by first clearing the AQ bit of the ASTAT register, then executing sixteen DIVQ instructions. The remainder is computed, after first setting MR2 to zero. This is necessary since MR1 automatically sign-extends into MR2. Also, the multiply must be executed with the unsigned switch. To ensure that the overflow flag is clear, ASTAT is set to zero before returning.

In both subroutines, the computation of the remainder requires only one extra cycle, so it is unlikely you would need to remove it for speed. If it is a problem to have the multiply registers altered, remove the multiply/subtract instruction just before the return, and remove the register transfers to MR0 and MR1 in the first two multifunction instructions. Be sure to remove the MR2=0; instruction in the *unsigned_div* subroutine also.

```
.MODULE/ROM   Divide_solution;

{
This module can be used to generate correct results when using the divide primitives
of the ADSP-2100 family. The code is organized in sections. This entire module can
be used to handle all error conditions, or individual sections can be removed to
increase execution speed.

  Entry Points
  signed_div Computes 16-bit signed quotient
  unsigned_div Computes 16-bit unsigned quotient

  Calling Parameters
  AX0 = 16-bit divisor
  AY0 = Lower 16 bits of dividend
  AY1 = Upper 16 bits of dividend
```

Division Exceptions   B

```
        Return Values
        AR = 16-bit quotient
        MR0 = 16-bit remainder
        AV flag set if divide would overflow

        Altered Registers
        AX0, AX1, AR, AF, AY0, AY1, MR, MY0

        Computation Time: 30 cycles
}

.ENTRY          signed_div, unsigned_div;

signed_div:     MR0=AY0,AF=AX0+AY1;         {Take divisor's absolute value}
                MR1=AY1, AR=ABS AX0;        {See if divisor, dividend have
                                             same magnitude}

test_1:         IF NE JUMP test_2;          {If divisor non-zero, do test 2}
                ASTAT=H#4;                  {Divide by zero, so overflow}
                RTS;                        {Return to calling program}

test_2:         IF NOT AV JUMP test_3;      {If divisor H#8000, then the}
                AY0=AY1, AF=ABS AY1;        {quotient is simply -AY1}
                IF NOT AV JUMP recover_sign;
                ASTAT=H#4;                  {H#8000 divided by H#8000,}
                RTS;                        {so overflow}

test_3:         AF=PASS AF;                 {Check for division overflow}
                IF NE JUMP test_4;          {Not equal, jump test 4}
                AY0=H#8000;                 {Quotient equals -1}
                ASTAT=H#0;                  {Clear AS bit of ASTAT}
                JUMP recover_sign;          {Compute remainder}

test_4:         AF=ABS AY1;                 {Get absolute of dividend}
                AR=ABS AX0;                 {Restore AS bit of ASTAT}
                AF=AF-AR;                   {Check for division overflow}
                IF LT JUMP do_divs;         {If Divisor>Dividend do divide}
                ASTAT=H#4;                  {Division overflow}
                RTS;
```

Listing B.1   Division Error Routine (continues on next page)

B Division Exceptions

```
do_divs:        DIVS AY1, AR; DIVQ AR;          {Compute sign of quotient}
                DIVQ AR; DIVQ AR;
                DIVQ AR; DIVQ AR;
                DIVQ AR; DIVQ AR;
                DIVQ AR; DIVQ AR;
                DIVQ AR; DIVQ AR;
                DIVQ AR; DIVQ AR;
                DIVQ AR; DIVQ AR;


recover_sign:   MY0=AX0,AR=PASS AY0;            {Put quotient into AR}
                IF NEG AR=-AY0;                 {Restore sign if necessary}
                MR=MR-AR*MY0 (SS);              {compute remainder dividend neg}
                RTS;                            {Return to calling program}


unsigned_div:   MR0=AY0, AF=PASS AY1;           {Move dividend MSW to AF}
                MR1=AY1, AR=PASS AX0;           {Is MSB set?}


test_10:        IF GT JUMP test_11;             {No, so check overflow}
                ASTAT=H#4;                      {Yes, so set overflow flag}
                RTS;                            {Return to caller}


test_11:        AR=AY1-AX0;                     {Is divisor<dividend?}
                IF LT JUMP do_divq;             {No, so go do unsigned divide}
                ASTAT=H#4;                      {Set overflow flag}
                RTS;


do_divq:        ASTAT=0;                        {Clear AQ flag}
                DIVQ AX0; DIVQ AX0;             {Do the divide}
                DIVQ AX0; DIVQ AX0;
                DIVQ AX0; DIVQ AX0;
                DIVQ AX0; DIVQ AX0;
                DIVQ AX0; DIVQ AX0;
                DIVQ AX0; DIVQ AX0;
                DIVQ AX0; DIVQ AX0;
                DIVQ AX0; DIVQ AX0;


uremainder:     MR2=0;                          {MR0 and MR1 previous set}
                MY0=AX0, AR=PASS AY0;           {Divisor in MXO, Quotient in AR}
                MR=MR-AR*MY0 (UU);              {Determine remainder}
                RTS;                            {Return to calling program}


.ENDMOD;
```

Listing B.1   Division Error Routine

Numeric Formats ■ C

C.1 OVERVIEW

ADSP-2100 family processors support 16-bit fixed-point data in hardware. Special features in the computation units allow you to support other formats in software. This appendix describes various aspects of the 16-bit data format. It also describes how to implement a block floating-point format in software.

C.2 UNSIGNED OR SIGNED: TWOS-COMPLEMENT FORMAT

Unsigned binary numbers may be thought of as positive, having nearly twice the magnitude of a signed number of the same length. The least significant words of multiple precision numbers are treated as unsigned numbers.

Signed numbers supported by the ADSP-2100 family are in twos-complement format. Signed-magnitude, ones-complement, BCD or excess-n formats are not supported.

C.3 INTEGER OR FRACTIONAL

The ADSP-2100 family supports both fractional and integer data formats, with the exception that the ADSP-2100 processor does not perform integer multiplication. In an integer, the radix point is assumed to lie to the right of the LSB, so that all magnitude bits have a weight of 1 or greater. This format is shown in Figure C.1, which can be found on the following page. Note that in twos-complement format, the sign bit has a negative weight.

C Numeric Formats

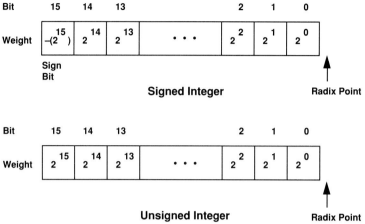

Figure C.1 Integer Format

In a fractional format, the assumed radix point lies within the number, so that some or all of the magnitude bits have a weight of less than 1. In the format shown in Figure C.2, the assumed radix point lies to the left of the 3 LSBs, and the bits have the weights indicated.

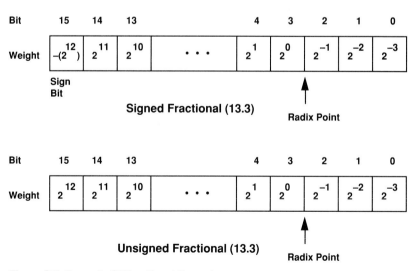

Figure C.2 Example Of Fractional Format

396

Numeric Formats   C

The notation used to describe a format consists two numbers separated by a period (.); the first number is the number of bits to the left of radix point, the second is the number of bits to the right of the radix point. For example, 16.0 format is an integer format; all bits lie to the left of the radix point. The format in Figure C.2 is 13.3.

Table C.1 shows the ranges of numbers representable in the fractional formats that are possible with 16 bits.

Format	Number of Integer Bits	Number of Fractional Bits	Largest Positive Value (0x7FFF) In Decimal	Largest Negative Value (0x8000) In Decimal	Value of 1 LSB (0x0001) In Decimal
1.15	1	15	0.999969482421875	−1.0	0.000030517578125
2.14	2	14	1.999938964843750	−2.0	0.000061035156250
3.13	3	13	3.999877929687500	−4.0	0.000122070312500
4.12	4	12	7.999755859375000	−8.0	0.000244140625000
5.11	5	11	15.999511718750000	−16.0	0.000488281250000
6.10	6	10	31.999023437500000	−32.0	0.000976562500000
7.9	7	9	63.998046875000000	−64.0	0.001953125000000
8.8	8	8	127.996093750000000	−128.0	0.003906250000000
9.7	9	7	255.992187500000000	−256.0	0.007812500000000
10.6	10	6	511.984375000000000	−512.0	0.015625000000000
11.5	11	5	1023.968750000000000	−1024.0	0.031250000000000
12.4	12	4	2047.937500000000000	−2048.0	0.062500000000000
13.3	13	3	4095.875000000000000	−4096.0	0.125000000000000
14.2	14	2	8191.750000000000000	−8192.0	0.250000000000000
15.1	15	1	16383.500000000000000	−16384.0	0.500000000000000
16.0	16	0	32767.000000000000000	−32768.0	1.000000000000000

Table C.1 Fractional Formats And Their Ranges

C.4   BINARY MULTIPLICATION

In addition and subtraction, both operands must be in the same format (signed or unsigned, radix point in the same location) and the result format is the same as the input format. Addition and subtraction are performed the same way whether the inputs are signed or unsigned.

In multiplication, however, the inputs can have different formats, and the result depends on their formats. The ADSP-2100 family assembly language allows you to specify whether the inputs are both signed, both unsigned, or one of each (mixed-mode). The location of the radix point in the result can be derived from its location in each of the inputs. This is

C   Numeric Formats

shown in Figure C.3. The product of two 16-bit numbers is a 32-bit number. If the inputs' formats are M.N and P.Q, the product has the format (M+P).(N+Q). For example, the product of two 13.3 numbers is a 26.6 number. The product of two 1.15 numbers is a 2.30 number.

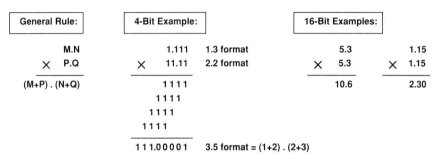

Figure C.3 Format Of Multiplier Result

C.4.1   Fractional Mode And Integer Mode

A product of 2 twos-complement numbers has two sign bits. Since one of these bits is redundant, you can shift the entire result left one bit. Additionally, if one of the inputs was a 1.15 number, the left shift causes the result to have the same format as the other input (with 16 bits of additional precision). For example, multiplying a 1.15 number by a 5.11 number yields a 6.26 number. When shifted left one bit, the result is a 5.27 number, or a 5.11 number plus 16 LSBs.

The ADSP-2100 family provides a mode (called the fractional mode) in which the multiplier result is always shifted left one bit before being written to the result register. (On the ADSP-2100 processor, this mode is always active; on other processors, the left shift can be omitted.) This left shift eliminates the extra sign bit when both operands are signed, yielding a correctly formatted result.

When both operands are in 1.15 format, the result is 2.30 (30 fractional bits). A left shift causes the multiplier result to be 1.31 which can be rounded to 1.15. Thus, if you use a fractional data format, it is most convenient to use the 1.15 format.

In the integer mode, the left shift does not occur. This is the mode to use if both operands are integers (in the 16.0 format). The 32-bit multiplier result is in 32.0 format, also an integer. On the ADSP-2100 only, the integer mode is not available; the 32.0 result gets shifted to 31.1 format. Because the MSB is still available in the 40-bit accumulator, a right shift can correct the result.

In all processors other than the ADSP-2100, fractional and integer modes are controlled by a bit in the MSTAT register. At reset, these processors default to the fractional mode, for compatibility with the ADSP-2100.

C.5   BLOCK FLOATING-POINT FORMAT

A block floating-point format enables a fixed-point processor to gain some of the increased dynamic range of a floating-point format without the overhead needed to do floating-point arithmetic. Some additional programming is required to maintain a block floating-point format, however.

A floating-point number has an exponent that indicates the position of the radix point in the actual value. In block floating-point format, a set (block) of data values share a common exponent. To convert a block of fixed-point values to block floating-point format, you would shift each value left by the same amount and store the shift value as the block exponent.

Typically, block floating-point format allows you to shift out non-significant MSBs, increasing the precision available in each value. You can also use block floating-point format to eliminate the possibility of a data value overflowing. Figure C.4 shows an example. The three data samples each have at least 2 non-significant, redundant sign bits. Each data value

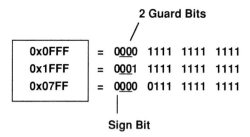

2 Guard Bits

0x0FFF	= 0000 1111 1111 1111
0x1FFF	= 0001 1111 1111 1111
0x07FF	= 0000 0111 1111 1111

Sign Bit

To detect bit growth into 2 guard bits, set SB=–2

Figure C.4 Data With Guard Bits

C Numeric Formats

can grow by these two bits (two orders of magnitude) before overflowing; thus, these bits are called *guard* bits. If it is known that a process will not cause any value to grow by more than these two bits, then the process can be run without loss of data. Afterward, however, the block must be adjusted to replace the guard bits before the next process.

Figure C.5 shows the data after processing but before adjustment. The block floating-point adjustment is performed as follows. Initially, the value of SB is –2, corresponding to the 2 guard bits. During processing, each resulting data value is inspected by the EXPADJ instruction, which counts the number of redundant sign bits and adjusts SB is if the number of redundant sign bits is less than 2. In this example, SB=–1 after processing, indicating that the block of data must be shifted right one bit to maintain the 2 guard bits. If SB were 0 after processing, the block would have to be shifted two bits right. In either case, the block exponent is updated to reflect the shift.

1. Check for Bit Growth

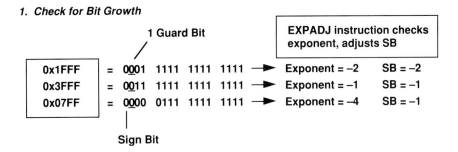

2. Shift Right to Restore Guard Bits

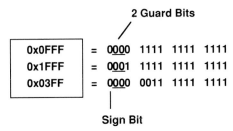

Figure C.5 Block Floating-Point Adjustment

Pin Descriptions ■ D

D.1 OVERVIEW

This appendix contains pin lists for the ADSP-2100, ADSP-2101, ADSP-2105, ADSP-2111 and ADSP-21msp50. The name and function of each pin are shown. Consult individual data sheets to find the pin configurations for specific packages.

D.2 ADSP-2100 PIN DESCRIPTION

Pin Name	Type	Function
PMA13-PMA0	Output	Address for program memory
PMD23-PMD0	I/O	Data for program memory
$\overline{PMS}$	Output	Program memory select
$\overline{PMRD}$	Output	Program memory read enable
$\overline{PMWR}$	Output	Program memory write enable
PMDA	Output	Program memory data access
DMA13-DMA0	Output	Address for data memory
DMD15-DMD0	I/O	Data for data memory
$\overline{DMS}$	Output	Data memory select
$\overline{DMRD}$	Output	Data memory read enable
$\overline{DMWR}$	Output	Data memory write enable
DMACK	Input	Data memory acknowledge
$\overline{RESET}$	Input	Processor reset
$\overline{IRQ3}$-$\overline{IRQ0}$	Input	External interrupt requests
$\overline{BR}$	Input	Bus request
$\overline{BG}$	Output	Bus grant
TRAP	Output	TRAP instruction indicator
$\overline{HALT}$	Input	Halt processor
CLKIN	Input	External clock
CLKOUT	Output	Processor clock
GND	Power	Ground
VDD	Power	Power Supply

D   Pin Descriptions

D.3   ADSP-2101 PIN DESCRIPTION

Pin Name	Type	Function
A13-A0	Output	Address for program, data and boot memory spaces
D23-D0	I/O	Data for program memory and data memory spaces. For program memory, D23-D0 are used. For data memory, only D23-D8 are used. For boot memory, D15-D8 are inputs, D23-D22 are outputs (used as 2 MSBs of address) and all others are unused.
RESET	Input	Processor reset
IRQ2	Input	External interrupt request #2
BR	Input	External bus request
BG	Output	External bus grant
PMS	Output	External program memory select
DMS	Output	External data memory select
BMS	Output	Boot memory select
RD	Output	External memory read enable
WR	Output	External memory write enable
MMAP	Input	Memory map select
CLKIN	Input	External clock or quartz crystal connection
XTAL	Output	Quartz crystal connection
CLKOUT	Output	Processor clock

Serial Port 0

TFS0	I/O	Transmit frame sync
RFS0	I/O	Receive frame sync
SCLK0	I/O	Programmable serial clock
DT0	Output	Data transmit
DR0	Input	Data receive

Serial Port 1

TFS1/IRQ1	I/O	Transmit frame sync/External interrupt request #1
RFS1/IRQ0	I/O	Receive frame sync/External interrupt request #0
SCLK1	I/O	Programmable serial clock
DT1/FO	Output	Data transmit/Flag out
DR1/FI	Input	Data receive/Flag in

GND	Power	Ground
VDD	Power	Power Supply

Pin Descriptions   D

D.4   ADSP-2105 PIN DESCRIPTION

Pin Name	Type	Function
A13-A0	Output	Address for program, data and boot memory spaces
D23-D0	I/O	Data for program memory and data memory spaces. For program memory, D23-D0 are used. For data memory, only D23-D8 are used. For boot memory, D15-D8 are inputs, D23-D22 are outputs (used as 2 MSBs of address) and all others are unused.
RESET	Input	Processor reset
IRQ2	Input	External interrupt request #2
BR	Input	External bus request
BG	Output	External bus grant
PMS	Output	External program memory select
DMS	Output	External data memory select
BMS	Output	Boot memory select
RD	Output	External memory read enable
WR	Output	External memory write enable
MMAP	Input	Memory map select
CLKIN	Input	External clock or quartz crystal connection
XTAL	Output	Quartz crystal connection
CLKOUT	Output	Processor clock

Serial Port 1

Pin Name	Type	Function
TFS1/IRQ1	I/O	Transmit frame sync/External interrupt request #1
RFS1/IRQ0	I/O	Receive frame sync/External interrupt request #0
SCLK1	I/O	Programmable serial clock
DT1/FO	Output	Data transmit/Flag out
DR1/FI	Input	Data receive/Flag in
GND	Power	Ground
VDD	Power	Power Supply

D   Pin Descriptions

D.5   ADSP-2111 PIN DESCRIPTION

Pin Name	Type	Function
A13-A0	Output	Address for program, data and boot memory spaces
D23-D0	I/O	Data for program memory and data memory spaces. For program memory, D23-D0 are used. For data memory, only D23-D8 are used. For boot memory, D15-D8 are inputs, D23-D22 are outputs (used as 2 MSBs of address) and all others are unused.
RESET	Input	Processor reset
IRQ2	Input	External interrupt request #2
BR	Input	External bus request
BG	Output	External bus grant
PMS	Output	External program memory select
DMS	Output	External data memory select
BMS	Output	Boot memory select
RD	Output	External memory read enable
WR	Output	External memory write enable
MMAP	Input	Memory map select
CLKIN	Input	External clock or quartz crystal connection
XTAL	Output	Quartz crystal connection
CLKOUT	Output	Processor clock

Host Interface Port

HSEL	Input	HIP select
HACK	Output	HIP acknowledge
HSIZE	Input	8/16 bit host select; 0=16-bit; 1=8-bit
BMODE	Input	Boot mode select; 0=normal (2101); 1=HIP
HMD0	Input	Bus strobe select; 0=RD, WR; 1=RW, DS
HMD1	Input	HIP address/data mode select; 0=separate; 1=multiplexed
HRD/HRW	Input	HIP Read strobe/Read/Write select
HWR/HDS	Input	HIP Write strobe/Host data strobe select
HD15-0/ HAD15-0	I/O	HIP Data/Data and Address
HA2/ALE	Input	Host address 2/Address Latch Enable
HA1-0/ unused	Input	Host addresses 1 and 0

Serial Port 0

TFS0	I/O	Transmit frame sync
RFS0	I/O	Receive frame sync
SCLK0	I/O	Programmable serial clock
DT0	Output	Data transmit
DR0	Input	Data receive

404

Pin Descriptions   D

Serial Port 1

TFS1/IRQ1	I/O	Transmit frame sync/External interrupt request #1
RFS1/IRQ0	I/O	Receive frame sync/External interrupt request #0
SCLK1	I/O	Programmable serial clock
DT1/FO	Output	Data transmit/Flag out
DR1/FI	Input	Data receive/Flag in
FL2-0	Output	General purpose Flag pins
GND	Power	Ground
VDD	Power	Power Supply

D.6   ADSP-21msp50 PIN DESCRIPTION

Pin Name	*Type*	*Function*
A13-A0	Output	Address for program, data and boot memory spaces
D23-D0	I/O	Data for program memory and data memory spaces. For program memory, D23-D0 are used. For data memory, only D23-D8 are used. For boot memory, D15-D8 are inputs, D23-D22 are outputs (used as 2 MSBs of address) and all others are unused.
RESET	Input	Processor reset
IRQ2	Input	External interrupt request #2
BR	Input	External bus request
BG	Output	External bus grant
PMS	Output	External program memory select
DMS	Output	External data memory select
BMS	Output	Boot memory select
RD	Output	External memory read enable
WR	Output	External memory write enable
MMAP	Input	Memory map select
CLKIN	Input	External clock or quartz crystal connection
XTAL	Output	Quartz crystal connection
CLKOUT	Output	Processor clock

(ADSP-21msp50 Pin Description continued on next page)

D Pin Descriptions

Host Interface Port

$\overline{\text{HSEL}}$	Input	HIP select
$\overline{\text{HACK}}$	Output	HIP acknowledge
HSIZE	Input	8/16 bit host select; 0=16-bit; 1=8-bit
BMODE	Input	Boot mode select; 0=normal (2101); 1=HIP
HMD0	Input	Bus strobe select; 0=$\overline{\text{RD}}$, $\overline{\text{WR}}$; 1=RW, $\overline{\text{DS}}$
HMD1	Input	HIP address/data mode select; 0=separate; 1=multiplexed
$\overline{\text{HRD}}$/HRW	Input	HIP Read strobe/Read/Write select
$\overline{\text{HWR}}$/$\overline{\text{HDS}}$	Input	HIP Write strobe/Host data strobe select
HD15-0/ HAD15-0	I/O	HIP Data/Data and Address
HA2/ALE	Input	Host address 2/Address Latch Enable
HA1-0/ unused	Input	Host addresses 1 and 0

Serial Port 0

TFS0	I/O	Transmit frame sync
RFS0	I/O	Receive frame sync
SCLK0	I/O	Programmable serial clock
DT0	Output	Data transmit
DR0	Input	Data receive

Serial Port 1

TFS1/$\overline{\text{IRQ1}}$	I/O	Transmit frame sync/External interrupt request #1
RFS1/$\overline{\text{IRQ0}}$	I/O	Receive frame sync/External interrupt request #0
SCLK1	I/O	Programmable serial clock
DT1/FO	Output	Data transmit/Flag out
DR1/FI	Input	Data receive/Flag in

406

Pin Descriptions   D

Analog Interface

VINNORM	Input	Inverting terminal of the NORM amplifier for the encoder section (ADC)
VFBNORM	Output	Terminal of the NORM amplifier for the encoder section
VINAUX	Input	Inverting terminal of the AUX amplifier for the encoder section (ADC)
VFBAUX	Output	Terminal of the AUX amplifier for the encoder section
VOUTP	Output	Non-inverting terminal of the output differential amplifier from the decoder (DAC)
VOUTN	Output	Inverting terminal of the output differential amplifier from the decoder
VREFOUT	Output	Reference voltage
FL2-0	Output	General purpose Flag pins
$\overline{\text{PWD}}$	Input	Powerdown processor
PWDACK	Output	Powerdown acknowledge
VDD	Power	Digital power supply
GND_D	Power	Digital ground
VCC	Power	Analog power supply
GND_A	Power	Analog ground

Control/Status Registers ■   E

E.1   OVERVIEW

This appendix shows bit definitions for 1) the memory-mapped control registers and 2) other control and status registers of all ADSP-21xx processors. The memory-mapped registers are listed in descending address order. Default bit values at reset are shown; if no value is shown, the bit is undefined at reset. Reserved bits are shown on a gray field. These bits should always be written with zeros.

Memory-Mapped Registers

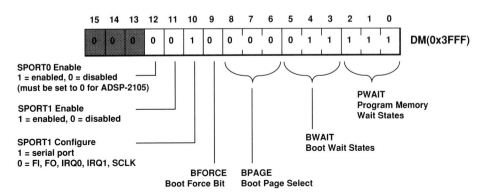

System Control Register   (ADSP-2101, ADSP-2105, ADSP-2111, ADSP-21msp50)

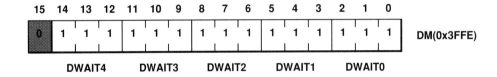

Data Memory Wait State Control Register   (ADSP-2101, ADSP-2105, ADSP-2111, ADSP-21msp50)

E Control/Status Registers

Memory-Mapped Registers

Timer Registers   (ADSP-2101, ADSP-2105, ADSP-2111, ADSP-21msp50)

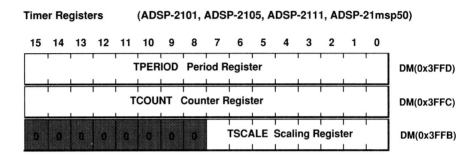

SPORT0 Multichannel Word Enable Register   (ADSP-2101, ADSP-2111, ADSP-21msp50)

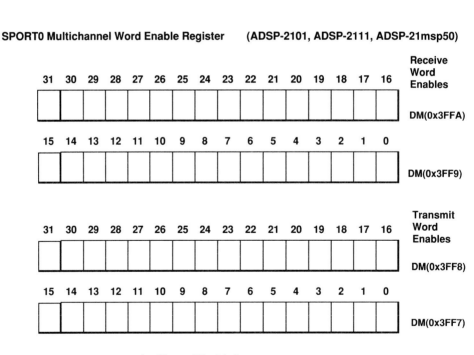

1 = Channel Enabled
0 = Channel Ignored

Control/Status Registers   E

Memory-Mapped Registers

SPORT0 Control Register   **(ADSP-2101, ADSP-2111, ADSP-21msp50)**

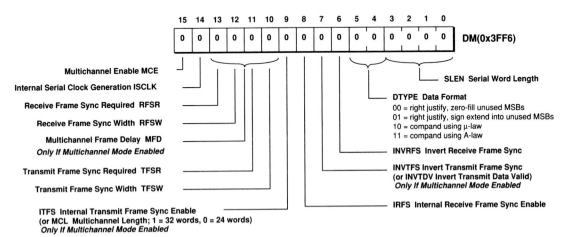

SPORT0 SCLKDIV   **(ADSP-2101, ADSP-2111, ADSP-21msp50)**
Serial Clock Divide Modulus

SPORT0 RFSDIV   **(ADSP-2101, ADSP-2111, ADSP-21msp50)**
Receive Frame Sync Divide Modulus

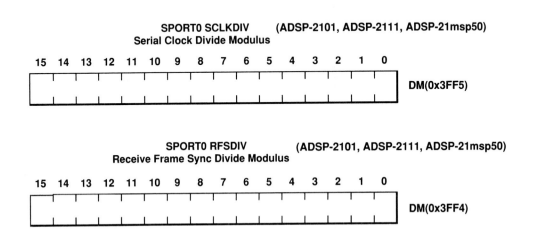

411

E   Control/Status Registers

Memory-Mapped Registers

SPORT0 Autobuffer Control Register   (ADSP-2101, ADSP-2111, ADSP-21msp50)

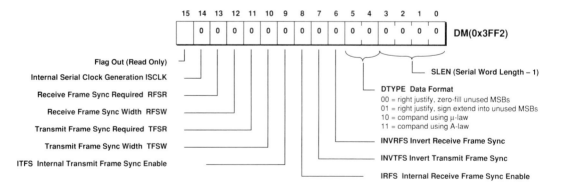

SPORT1 Control Register   (ADSP-2101, ADSP-2105, ADSP-2111, ADSP-21msp50)

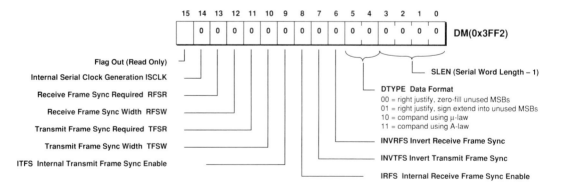

Control/Status Registers   E

Memory-Mapped Registers

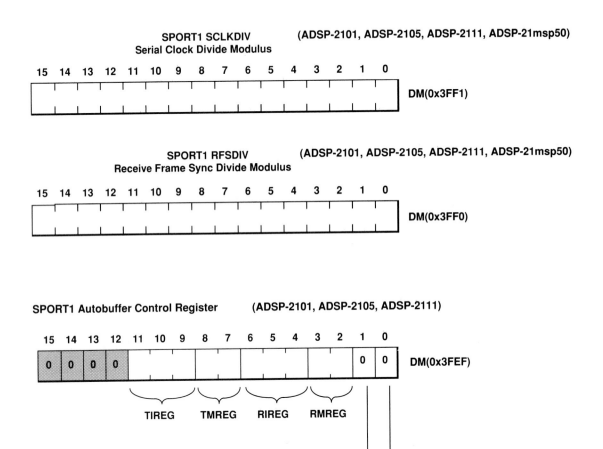

SPORT1 SCLKDIV
Serial Clock Divide Modulus   (ADSP-2101, ADSP-2105, ADSP-2111, ADSP-21msp50)

| 15 | 14 | 13 | 12 | 11 | 10 | 9 | 8 | 7 | 6 | 5 | 4 | 3 | 2 | 1 | 0 |

DM(0x3FF1)

SPORT1 RFSDIV
Receive Frame Sync Divide Modulus   (ADSP-2101, ADSP-2105, ADSP-2111, ADSP-21msp50)

| 15 | 14 | 13 | 12 | 11 | 10 | 9 | 8 | 7 | 6 | 5 | 4 | 3 | 2 | 1 | 0 |

DM(0x3FF0)

SPORT1 Autobuffer Control Register   (ADSP-2101, ADSP-2105, ADSP-2111)

| 15 | 14 | 13 | 12 | 11 | 10 | 9 | 8 | 7 | 6 | 5 | 4 | 3 | 2 | 1 | 0 |
| 0 | 0 | 0 | 0 | | | | | | | | | | | 0 | 0 |

DM(0x3FEF)

TIREG   TMREG   RIREG   RMREG

TBUF Transmit Autobuffering Enable
RBUF Receive Autobuffering Enable

E Control/Status Registers

Memory-Mapped Registers

Analog Autobuffer/Powerdown Control Register   (ADSP-21msp50)

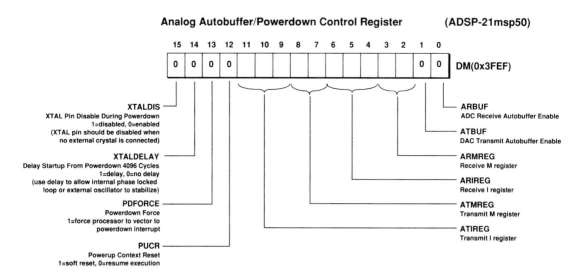

15	14	13	12	11	10	9	8	7	6	5	4	3	2	1	0
0	0	0	0											0	0

DM(0x3FEF)

XTALDIS
XTAL Pin Disable During Powerdown
1=disabled, 0=enabled
(XTAL pin should be disabled when
no external crystal is connected)

XTALDELAY
Delay Startup From Powerdown 4096 Cycles
1=delay, 0=no delay
(use delay to allow internal phase locked
loop or external oscillator to stabilize)

PDFORCE
Powerdown Force
1=force processor to vector to
powerdown interrupt

PUCR
Powerup Context Reset
1=soft reset, 0=resume execution

ARBUF
ADC Receive Autobuffer Enable

ATBUF
DAC Transmit Autobuffer Enable

ARMREG
Receive M register

ARIREG
Receive I register

ATMREG
Transmit M register

ATIREG
Transmit I register

Analog Control Register   (ADSP-21msp50)

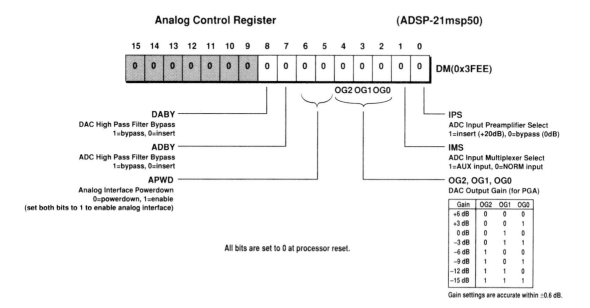

15	14	13	12	11	10	9	8	7	6	5	4	3	2	1	0
0	0	0	0	0	0	0	0	0	0	0	0	0	0	0	0

OG2 OG1 OG0

DM(0x3FEE)

DABY
DAC High Pass Filter Bypass
1=bypass, 0=insert

ADBY
ADC High Pass Filter Bypass
1=bypass, 0=insert

APWD
Analog Interface Powerdown
0=powerdown, 1=enable
(set both bits to 1 to enable analog interface)

IPS
ADC Input Preamplifier Select
1=insert (+20dB), 0=bypass (0dB)

IMS
ADC Input Multiplexer Select
1=AUX input, 0=NORM input

OG2, OG1, OG0
DAC Output Gain (for PGA)

Gain	OG2	OG1	OG0
+6 dB	0	0	0
+3 dB	0	0	1
0 dB	0	1	0
–3 dB	0	1	1
–6 dB	1	0	0
–9 dB	1	0	1
–12 dB	1	1	0
–15 dB	1	1	1

Gain settings are accurate within ±0.6 dB.

All bits are set to 0 at processor reset.

414

Control/Status Registers   E

Memory-Mapped Registers

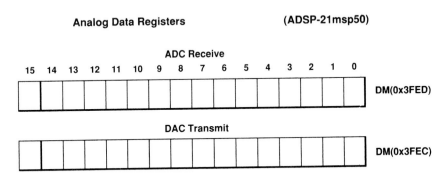

Analog Data Registers   (ADSP-21msp50)

ADC Receive

| 15 | 14 | 13 | 12 | 11 | 10 | 9 | 8 | 7 | 6 | 5 | 4 | 3 | 2 | 1 | 0 |

DM(0x3FED)

DAC Transmit

DM(0x3FEC)

No registers are located between 0x3FEC and 0x3FE8.

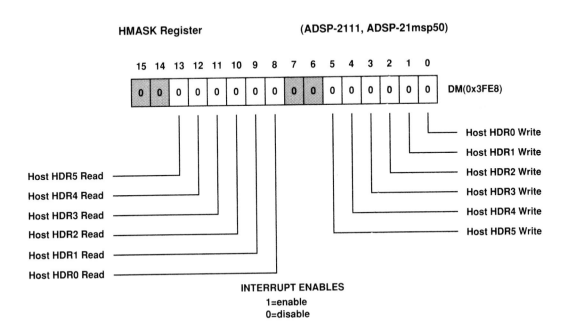

HMASK Register   (ADSP-2111, ADSP-21msp50)

| 15 | 14 | 13 | 12 | 11 | 10 | 9 | 8 | 7 | 6 | 5 | 4 | 3 | 2 | 1 | 0 |
| 0 | 0 | 0 | 0 | 0 | 0 | 0 | 0 | 0 | 0 | 0 | 0 | 0 | 0 | 0 | 0 |

DM(0x3FE8)

Host HDR0 Write
Host HDR1 Write
Host HDR2 Write
Host HDR3 Write
Host HDR4 Write
Host HDR5 Write

Host HDR5 Read
Host HDR4 Read
Host HDR3 Read
Host HDR2 Read
Host HDR1 Read
Host HDR0 Read

INTERRUPT ENABLES
1=enable
0=disable

E Control/Status Registers

Memory-Mapped Registers

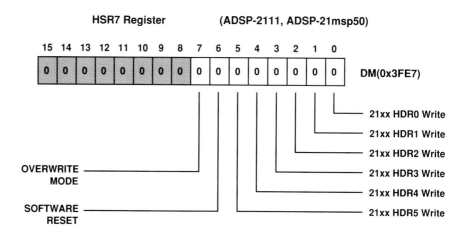

HSR7 Register   (ADSP-2111, ADSP-21msp50)

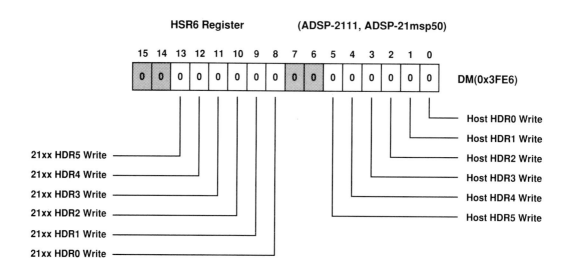

HSR6 Register   (ADSP-2111, ADSP-21msp50)

Control/Status Registers   E

Memory-Mapped Registers

HIP Data Registers　　　　(ADSP-2111, ADSP-21msp50)

HDR5

15	14	13	12	11	10	9	8	7	6	5	4	3	2	1	0

DM(0x3FE5)

HDR4

DM(0x3FE4)

HDR3

DM(0x3FE3)

HDR2

DM(0x3FE2)

HDR1

DM(0x3FE1)

HDR0

DM(0x3FE0)

E Control/Status Registers

Non-Memory-Mapped Registers

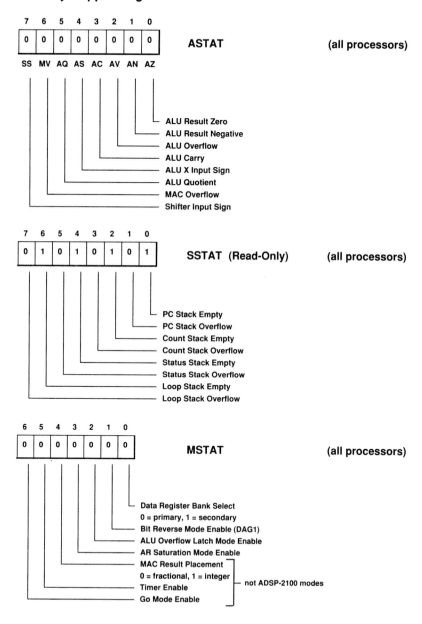

7	6	5	4	3	2	1	0
0	0	0	0	0	0	0	0
SS	MV	AQ	AS	AC	AV	AN	AZ

ASTAT (all processors)

- ALU Result Zero
- ALU Result Negative
- ALU Overflow
- ALU Carry
- ALU X Input Sign
- ALU Quotient
- MAC Overflow
- Shifter Input Sign

7	6	5	4	3	2	1	0
0	1	0	1	0	1	0	1

SSTAT (Read-Only) (all processors)

- PC Stack Empty
- PC Stack Overflow
- Count Stack Empty
- Count Stack Overflow
- Status Stack Empty
- Status Stack Overflow
- Loop Stack Empty
- Loop Stack Overflow

6	5	4	3	2	1	0
0	0	0	0	0	0	0

MSTAT (all processors)

- Data Register Bank Select
  0 = primary, 1 = secondary
- Bit Reverse Mode Enable (DAG1)
- ALU Overflow Latch Mode Enable
- AR Saturation Mode Enable
- MAC Result Placement
  0 = fractional, 1 = integer
- Timer Enable
- Go Mode Enable

not ADSP-2100 modes

418

Control/Status Registers   E

Non-Memory-Mapped Registers

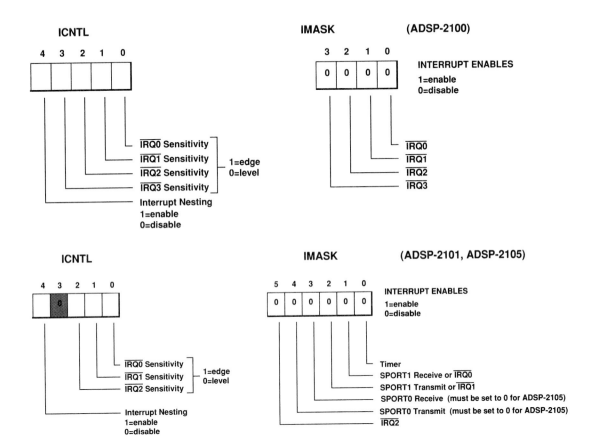

E   Control/Status Registers

Non-Memory-Mapped Registers

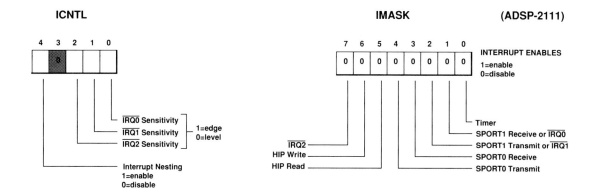

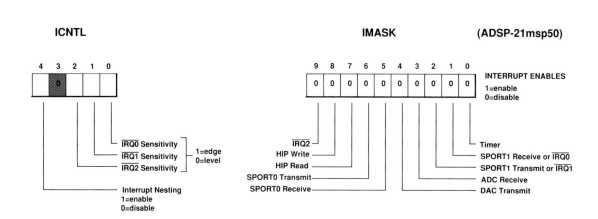

Control/Status Registers   E

Non-Memory-Mapped Registers

IFC (Write-Only)   (ADSP-2101, ADSP-2105, ADSP-2111)

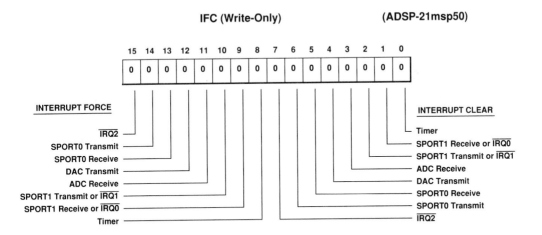

IFC (Write-Only)   (ADSP-21msp50)

Index ■

423

Index

Index

Index

426

Index

Index

Index